Political Nature

Political Nature

Environmentalism and the Interpretation of
Western Thought

John M. Meyer

The MIT Press
Cambridge, Massachusetts
London, England

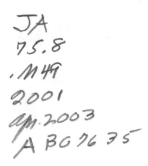

This book was set in Sabon by Achorn Graphic Services, Inc., on the Miles 33 system.

Printed and bound in the United States of America.

Library of Congress Cataloging-in-Publication Data

Meyer, John M.
    Political nature : environmentalism and the interpretation of Western
    thought / John M. Meyer.
        p.   cm.
    Includes bibliographical references and index.
    ISBN 0-262-13390-3 (hc. : alk. paper) — ISBN 0-262-63224-1 (pbk. : alk.
    paper)
    1. Political ecology. 2. Environmentalism. I. Title.
 JA75.8 .M49 2001
  320.5—dc21

                                                              2001030459

To Carolyn

Contents

Preface

I worked as a political organizer for the six years prior to my graduate studies, often addressing environmental issues. Among the necessities of this activism are the obligation to develop strategies, to respond to opposition quickly, to shore up support for proposals, and to maintain one's own confidence—and the confidence of others—in the face of a daily whirlwind of activity. Broader insight into where we are going and where we want to go can be fostered in this atmosphere, but I also found that it can be stifled by the lack of opportunity for reflection.

I reentered the academy in search of this opportunity and with a sense that our human relationship with the natural environment is of fundamental importance to our social and political values. I immersed myself in the study of both political theory and environmental ideas. As I read historical and contemporary literature in political theory, I searched for insight into human-environment relations. In one sense, not surprisingly, I found little. Environmental concerns as we understand them today have not played a major role in Western political thought. In another sense, however, it became increasingly apparent that the argumentative strategies by which various theorists have sought to relate ideas about "nature" to their political vision were of great relevance. I found myself prodding these connections to better understand the relationship. While often I did not embrace these views, I discovered in them the source for reflection that I had sought.

My initial examination of writings by contemporary environmentalist thinkers prompted a different response. Here I often shared sentiments and underlying concerns, yet found journals such as *Environmental Ethics* and many books in the field to be weighed down with attempts to define nature as intrinsically valuable and with arguments for the transformation of our worldview. Quite frankly, I found these to be both theoretically and practically unpersuasive. Let me reserve my theoretical response for later in the book, but touch briefly upon some practical reservations here. My experiences as a political organizer left me with an appreciation for the widespread good sense of a great many people, both in general and with particular regard to environmental concerns. While they often expressed frustration with economic and political elites or skepticism regarding the likelihood for positive change, I only occasionally encountered either hostility or apathy. The environmental concern I found was rooted in diverse life experiences and values.

The challenge seemed to me then, and now, to figure out how this widely shared concern might become much more integral to our social, economic, and political decision-making. However, the focus on intrinsic value and a transformation of worldview seems based on a rejection of the convictions of most people and a rather naïve assumption that attention to political ideas and institutions is at best of secondary importance. As such, it is rather tangential to the challenge at hand. Over the years that I have been working on this book, this disappointment with the environmental literature has increasingly been voiced by others. Thus I hope that my analysis here will be seen as a contribution to a growing dialogue on the relationship between nature and political order.

Acknowledgments

I wish to thank the many friends, colleagues, teachers, and correspondents who contributed to this book. Charles Anderson, John Barry, Mark Brown, Michael Goodhart, Harvey Jacobs, Brian Kroeger, Andrew Murphy, Sam Nelson, Patrick Riley, David Siemers, Joe Soss, Greg Streich, Tracy Wahl, Harlan Wilson, and Julie White each read a portion of the manuscript and took the time to critically discuss my ideas. Thanks, also, to Peter Cannavò, Booth Fowler, and several reviewers, who read the whole manuscript in some form and helped enormously as I worked to refine and clarify my argument. I especially wish to express my appreciation to Marion Smiley and Bernard Yack, who not only read and commented upon the manuscript but whose intellectual guidance was essential to whatever insight I am able to offer here.

If I had incorporated all of the judicious advice that I received, the resulting book surely would have been better, although it may never have been completed. I hope that the book is suggestive enough to encourage others to reflect upon themes left underdeveloped here.

A version of chapter 3 was published as "Interpreting Nature and Politics in the History of Western Thought: The Environmentalist Challenge" in *Environmental Politics*, vol. 8, no. 2, pp. 1–23, Frank Cass Publishers, London. Thanks to the publisher for allowing me to revise and reprint it here.

At a key stage in this project, Stephen Levine of Victoria University of Wellington, New Zealand offered an unexpected and irresistible visiting appointment. There, the beautiful country combined with the support of Stephen, Pat Moloney, and the other members of the Department of Politics to provide an unbeatable setting in which to complete the doctoral dissertation from which this book emerged. In our present home on California's north coast, I wish to thank the students and colleagues at Humboldt who have made me feel welcome. Thanks, also, to the HSU Foundation, which provided a small amount of financial assistance at the final stages of my work on this book.

I have dedicated this book to Carolyn Benson because without her unwavering confidence, encouragement, empathy, and love, it would have been impossible and my life would be unimaginable. She has taught me most, about the most important things. The lives of our children, Jacob and Emelia, have forever been entwined with my work on this project. Now that it is done, I hope to repay their indulgence by taking more time with them to experience the magnificent place in which we now live. Finally, I wish to acknowledge a lifetime of loving support from my parents. Mom and Dad have always believed in me and urged me to do "whatever makes you happy." Their supreme confidence in my abilities and aspirations is a gift that has brought me happiness, even when work on this book has put me in less pleasant moods.

Political Nature

1

Introduction

How should human communities be ordered? Those who engage in politics—in its broadest sense—seek to address this question. What is and what should be the relationship between the order of human communities and the order of nature? This is one of the monumental questions confronted—sometimes explicitly, always implicitly—by political thinkers over the millennia. It is also a question that contemporary environmentalists regularly seek to address. Thus the question is not only a perennial one, it is timely.

Among environmentalist thinkers, calls for a new "ecological worldview" have been commonplace. These are typically framed as an appeal to nature from which economic, social, and political principles follow. Once we properly identify the natural standard, this argument suggests, we can organize human communities in an appropriate manner. The "laws of nature," one environmentalist author contends, offer us "directions . . . for human settlements and systems."[1] "Certain outlooks on politics and public policy flow naturally" from an "ecological consciousness," another widely read work insists.[2] These sorts of claims are troubling for precisely the same reason that they are powerful—they cede enormous authority over human affairs to something deemed "nature." At the very least, any *democratic* politics would be imperiled by an acceptance of these claims. I argue that this elevation of nature to the level of directive ultimately is rendered incoherent by its failure to recognize *any* meaningful role for politics.

Very recently, some of the most perceptive environmentalist thinkers have pointed out the dangers of appealing to nature in this manner and have been more circumspect, distancing themselves from such a relationship. Yet the situation that motivated this appeal—the often devastating effects of human activity upon our environment—continues unabated. Is there a way out of this conundrum? Is it possible to relate nature to politics without presenting nature as an authority? I believe that it is both possible and necessary. However, it is not easy. It is even harder than the critics often recognize or acknowledge. One key reason for the difficulty is that the actual connection between nature and politics has been gravely misrepresented by the ways that both environmentalist thinkers and many other scholars have understood and discussed the history of Western thought.

Interpreters of the nature-politics relationship in the history of Western thought routinely portray it in one of two mutually exclusive yet equally misleading ways. On the one hand, a number argue that the distinguishing characteristic of Western thought is that politics (and human culture generally) is completely divorced from nature. I refer to this as the *dualist* account. Some of these interpreters identify the emergence of seventeenth-century social contract philosophy, with its emphasis on the non-natural quality of political agreement, as the origin of a broader nature-culture dualism in the modern West. Some argue for a dualism traced either to the origin of Christianity or to the ancient Greek origin of Western philosophy itself. On the other hand, many view Western political thought as replete with normative theories derived from conceptions of nature, whether that conception be the teleology of Aristotelians, the clocklike mechanism of early modern scientists, or the invisible hand of Darwinian selection. I refer to this as the *derivative* interpretation.

In the chapters that follow, I closely examine the nature–politics relationship in the writings of environmentalist thinkers and in the writings of two key Western philosophers of nature and politics: Thomas Hobbes and Aristotle. I argue that the dominant interpretations—dualist and derivative—fail to take into account the interactions between our conceptions of nature and politics. Only when we overcome the limitations of these interpretations can we speak about nature and politics together. Unless we are able to speak in this way, our theorizing will be handicapped either by an inability to appreciate the inescapable significance of nature and the natural world for contemporary politics, or by a failure to acknowledge the crucial role of political conceptions and judgments in shaping the ways in which nature is understood. In either case, the tragic consequence is that our actions are left unguided at a time when guidance is desperately needed.

A brief discussion of environmental concerns may illustrate why theoretical guidance is now essential and may explain my objective in the chapters that follow. These concerns have become important political issues in recent decades as the magnitude of many environmental threats has increased. At the same time, however, environmental policies often have failed to live up to their initial promise. For many environmentalist thinkers, the result has been disillusionment with "technocratic fixes," "incremental policy-making," and "politics as usual" as solutions to environmental problems. "Reform environmentalism"—both the practical efforts of such mainstream organizations as the Sierra Club and the National Wildlife Federation and the philosophy that underpins these efforts—has been deemed by critics to be too accommodating to a dominant and destructive attitude toward nature. As a consequence, these critics contend, "reformers" are ineffective in achieving even their own limited objectives.[3]

The loss of species diversity, ozone depletion, the health effects of toxic discharges, and the likelihood of global warming, among other problems, should reinforce our appreciation for ecological arguments about the interdependence of life on earth.[4] Our relationship to the places in which we live is highlighted when we embrace these arguments. As a result, issues that were previously deemed peripheral to politics—and often appeared to be outside the public domain—now frequently appear central. This prompts us to reexamine established thinking about society and politics, an effort that ought to be the core of what some now call green political theory.[5] Yet as I have already noted, environmentalist thinkers as a rule rely more narrowly upon a perceived need to adopt a new conception of nature:

there is often a strong sense in which the natural world is taken as a model for the human world, and many of ecologism's prescriptions for political and social arrangements are derived from a particular view of how nature "is". This view—not surprisingly—is an ecological view.[6]

The crucial presumption here is that there are no significant factors intervening between our conception of nature and our social practices. Our most fundamental concern in this regard certainly should be with the ability of our political and social institutions to address environmental problems effectively. If so, then a single-minded emphasis on an ecological view of nature could only make sense if it were true that "political and social arrangements are derived from" it. However, our conception of politics itself intervenes here.

Politics is more than the sum of policies and practices; it first must be conceptualized in some particular manner. For example, is politics envisioned as potentially deliberative and open to a wide array of values? Then it is likely to be regarded as a feasible mechanism to address environmental concerns.[7] By contrast, those who regard politics and the state as rigid, inflexible, and unitary are likely to view it as unresponsive to these concerns. Adhering to such a view, some libertarian thinkers have asserted that it is the "free market"—understood as an evolutionary, self-organizing phenomenon—that can best incorporate new environmental concerns and values.[8] Others have pointed to "civil society" as providing a more amenable home for action affecting environmental concerns.[9] One's political conception thus plays an essential role in the formulation of an argument about what humanity will do in response to environmental concerns. This role is one that I emphasize throughout my analysis of both past and present efforts to theorize about the nature–politics relationship.

In this book I ask: What are the implications of concluding that an ecological conception of nature should be incorporated into our social and political thought? It is in the answers that are suggested or offered to this question (and not the conclusion itself) that I am often critical of existing theorizing because I find that our political conceptions inevitably color the answers that we offer. Thus it is in the effort to answer this question that I am drawn to the broad and perplexing question of the relationship between conceptions of nature and of politics generally.

Rethinking the Nature–Politics Relationship

This book is both critical and reconstructive. I direct the critical argument against the two distinct perspectives, which I have termed dualism and derivation, common among both contemporary environmentalist thinkers and many interpreters of Hobbes, Aristotle, and other Western political theorists. While I have selected these labels to clearly specify the nature–politics relationship that is central to each interpretation, dualism and derivation are more often, albeit less precisely, identified as "humanism" and "naturalism," respectively. Whereas these latter terms are often used to describe conflicting traditions within the West, I wish to emphasize the sense in which dualism and derivation are each offered as a singular reading of the dominant perspective in Western thought.

Derivative interpretations often focus on the centrality of "natural right" within Western political thought.[10] Here it is argued that some directive or first principle has been the basis for discussions about the properly ordered human community. The most influential principles are those that appeal to the unchanging character of the cosmos. Of course, our understanding of the cosmos—of nature—has undergone dramatic change. As a consequence, the derivative interpretation seeks to describe the changes in political thought that have attended such milestones as the adoption of the so-called "modern" conception of nature in the seventeenth century.

Dualist interpretations read the same record of Western thought very differently. Here, the predominant story is one of communities that sought to tear themselves away from the cycles of nature by celebrating and elevating qualities believed to be distinctively human. Past Western thought is thus said to be characterized by an unbridgeable chasm between conceptions of nature on the one hand and of humanity and politics on the other.

Among environmentalists, the dualist interpretation is a basis for rejecting Western thought as inimical to a recognition of ecological interconnectedness. By contrast, the derivative interpretation becomes the basis for arguing that an *ecological* conception of nature must replace the flawed and outdated conceptions that until now have served as the basis of political authority. Thus, while the dualist interpretation is the mirror image of the argument that politics has been derived from some natural standard, the prescription offered to environmentalists by both schools of interpretation is the same. Both readings of the past reinforce the appeal to nature as the standard for a future ecological polity.

In an oft-cited observation, Raymond Williams convincingly asserted that "Nature is perhaps the most complex word in the language."[11] The reconstructive effort that I pursue here requires that I disentangle several distinct ways of speaking about nature and the role that it plays in human affairs. These distinctions are often elided in the interpretations that I criticize, so a brief outline of the key usages here will provide an orientation for my analysis.

Nature is sometimes defined as "all that exists" or the "collective phenomena of the world or universe."[12] However, before it can be meaningfully understood, it must somehow be distinguished from "everything." Typically, a quality or essence is identified by which we can portray nature as being more consequential. We will encounter several candidates for the quality that is "nature" in the course of this book, yet Williams is correct to argue that in each case, nature becomes "singular and abstracted."[13] Such a quality is then an idea or *conception* of nature.

This conception of nature does not commit one to a particular role for it within normative political theory. However, the familiarity of the directive role is so great that the transition is often ignored. By defining nature as a distinctive essence or principle, we appear to invite "propositions of the form . . . 'Nature shows . . .' or 'Nature teaches. . .' "[14] Rather than a mere effort at description, the capitalized idea of "Nature" is transformed here into a source of authority that can direct human affairs.

There is, of course, much disagreement about this appeal to nature (especially outside the community of environmentalist thinkers), which I explore in my analysis of dualist and derivative interpretations. What is often missing in the conflict between those who would embrace and those who would reject the appeal, however, is recognition of the distinctive quality of this role for nature. Discussing the relationship between nature and politics in a different way first requires us to draw upon an alternative understanding of the role that nature can play.

The role that must be recognized is a *constitutive* one. By doing so, we acknowledge the ways in which humans are inescapably natural beings, whose thought, actions, and potentialities are inextricably interdependent with and embedded within the world. Especially among environmentalists, it should be clear, this argument about human embeddedness in nature is hardly new. It is precisely the argument that is used to counter the dualism often characterized as central to Western thought. My most distinctive contribution to these discussions, therefore, does not lie in advancing this argument itself. It is by differentiating it from nature viewed as a source of direction or authority that I believe we can promote meaningful discussion.

This constitutive role for nature shapes our understanding of human experience in a multitude of ways. It is also key to another distinct usage of "nature" that is prominent in discussions of environmentalist concerns. This is a reference to "nature as a physical place," that can have meaning and value.[15] This, as Kate Soper notes, is "the nature we have destroyed and polluted and are asked to conserve and preserve."[16] Yet in this experiential sense, nature is distinct from the "singular and abstracted" quality described earlier. Williams explains the relationship between these two "natures":

we have here a case of a definition of quality [Nature] which becomes, through real usage, based on certain assumptions, a description of the world. . . . A singular name for the real multiplicity of things and living processes.[17]

The singular conception of nature can play the role of normative authority and can also be the basis for describing and understanding the world we inhabit.

Here, however, we encounter ambiguity. If all of the world qualified as nature, then it would no longer be a term of distinction and we would return to the uninformative definition with which we began. Yet anything less cannot be defined purely as a consequence of a conception of nature. When we speak of nature as a condition, a place, or a realm of experience—something familiar among, but by no means unique to, environmental discussions—we are creating a category whose boundaries are not authoritatively defined by a conception of nature and which are necessarily subjective and political. What we have is a category that might be described as a *political nature*. Not a unique product of either a conception of nature or of politics, it emerges from a dialectic between them.

There are, then, two distinct yet related threads to my discussions of nature in this book. The first focuses on the conceptions of nature held by various thinkers. What is the essence or principle identified as natural? Which principles play a major role in particular theories? The second focuses on the role played by these conceptions of nature. Thus the relationship between them and ideas about the ordering of human affairs becomes central. Here we must distinguish a constitutive from a directive role for nature. We must delineate the ways in which a dialectic between a conception of nature and politics creates the category that we experience as nature and shapes the social and political practices by which we address or fail to address the concerns generated through this experience. Central to my analysis is the argument that an understanding of this relationship is as or more important than the conception of nature itself.

This reconstruction of the relationship between conceptions of nature and politics emerges through my consideration of earlier presentations of it in the history of Western thought. While my first concern is to promote fresh thinking on environmental arguments, my analysis also addresses some persistent dilemmas of interpretation in Western political thought. The prevalence of derivative and dualist interpretations often leads to contradictory views of a single thinker, both of which are misguided. These can be found among environmentalists and also among political theorists uninterested in environmental questions. By focusing on how both of these interpretations play out in relation to the writings of Aristotle and Thomas Hobbes, I am able to make better sense of a number of arguments that each of these two pivotal thinkers advances.

Past thinkers have articulated a wide variety of conceptions of nature. Even if we limit our scope to the history of Western thought, this diversity remains great.[18] Nonetheless, two conceptions are the most historically prominent and influential in Western views of nature and the natural world. The first of these, often labeled "teleological" and sometimes "organic," is most closely associated with the thought of Aristotle, and was incorporated within a Christian framework by medieval Aristotelians. The second arose in large part as a reaction against the first. It is closely identified with the scientific revolution in seventeenth-century Europe and the rise of modern science that was its consequence. Labeled the "mechanistic" or the "modern" view of nature, it emerged from the works of such notable thinkers as Galileo, Descartes, and Newton.[19]

Aristotle and Hobbes each developed their political philosophy in relation to one of these two especially prominent conceptions of nature. Moreover, each elaborated their political philosophy and their natural philosophy in significant depth. As a consequence, their thought offers especially fertile grounds upon which to explore the conceptual relationship between nature and politics. Despite this, few interpreters begin with a serious analysis of their texts in both political and natural philosophy: Aristotle's *Physics* as well as his *Politics*, Hobbes's *De Corpore* as well as his *Leviathan*. By considering these and other works here, I am able to make sense of the relationship that exists between their conceptions of nature and politics, even where their intentions are less than clear.

Ironically, difficulties and apparent contradictions in interpretations of Hobbes and Aristotle are common precisely because their understandings of nature and politics are so thoroughly developed. The more fully one develops these, the more difficult it becomes to maintain the position that either a derivative or a dualist relationship exists between them. By disentangling constitutive from directive roles for a conception of nature, we can recognize that both Hobbes and Aristotle are best understood as advancing the former rather than the latter. This recognition allows me to present the constitutive role as the basis for a mutually determining—or dialectical—relationship between nature and politics. It also enables me to overcome some of the most difficult and seemingly contradictory aspects of Hobbes's and Aristotle's views on nature and politics, thus presenting a more intelligible reading of these two philosophers.

Only after we have disentangled a constitutive from a directive role for nature can we recognize that it is the former that is supported by the weight of evidence. Then it becomes possible to reject a view of politics as derivative of nature, replacing it with a dialectical relationship between the two. By doing so, we allow for the profoundly important yet (once stated) eminently commonsensical observation that a diversity of *attitudes* toward nature can be—and frequently is—present both within the philosophical thought and within the broader culture of a particular period.

Crucially, the existence of diverse attitudes toward nature within a common culture seems necessary to explain the growth of environmentalism itself. After all, this growth cannot be explained by the same attitude that environmentalist thinkers use to explain our currently destructive behavior. Listen to one recent effort to accommodate these two views: In a manner that will soon sound familiar, Rupert Sheldrake traces the "historical roots" of environmental destruction to the "mechanistic theory of nature that has dominated scientific thinking since the seventeenth century." By contrast, he makes the following claim on the same page:

Despite all this, a vague sense of the sacredness of nature, an unarticulated nostalgia, persists in many of us. The widespread desire to get back to nature, the need to find inspiration in the countryside or untamed wilderness, stems from this residual . . . sense . . . [It] can be categorized as "poetic," "romantic," "aesthetic" or "mystical." But as such [it] can only be part of our private lives.[20]

Sheldrake, like many others, seeks to emphasize the power and importance of these "poetic," "aesthetic," and "romantic" attitudes toward nature. However, given his commitment to an argument about the centrality of a mechanistic worldview, he is compelled to describe this alternative contemporary attitude toward nature as a "residual" that inexplicably "persists" from an age prior to the development of the mechanistic view in the seventeenth century.

If, while continuing to embrace a constitutive role for nature, we reject the presumption that a single directive attitude must be integral to it, then we may recognize that diverse and contradictory attitudes can and do coexist within the thought of a particular place or time. A "romantic" attitude toward nature no longer must be treated as a residual—it can be recognized as a consequential and important attitude held by many people today. We can acknowledge it as a significant source of motivation for many environmentalists. If so, however, then the final sentence in the quotation from Sheldrake becomes the most interesting. Why might it be the case that this particular attitude toward nature "can only be part of our private lives"? I suggest that this question can be fruitfully considered as an instance of a broader question that I explore in this book: Since there are divergent attitudes toward nature within a society, then how is it that certain attitudes become dominant in public decision-making, while others are seemingly relegated to the private sphere? This question can only be answered if we first have an appreciation for the significant role that a conception of politics can play in selecting particular values and attitudes as dominant or not. It is this appreciation that a dialectical relationship can highlight.

Interpretative Approach

Since the argument in this book emerges through my reading of the works of Hobbes, Aristotle, and environmentalist philosophy, I wish to say a few words about my method of interpreting texts. My concern is to understand how the key conceptual categories of nature and politics are related within the works that I consider. Exploring this relationship within a text is not necessarily the same as asking what is either claimed or intended by the author in question.[21] Indeed, in some works considered later, I argue that the author's claims are contrary to the nature–politics relationship actually present within the theory discussed. Hobbes, for example, is quite explicit at a number of points in asserting that an appropriate political philosophy is one derived from a true conception of nature. In exploring Hobbes's texts, however, I find this derivative relationship to be unsustainable. In its place, I sketch an interactive relationship between nature and politics that is apparent upon a close examination of his arguments.[22]

The potential for the success of the interpretive approach that I utilize—exploring a text for insights that an author may not have explicated—depends in large part upon the texts with which one begins. Works developed with care and in depth, with a richness of argument, are those in which such an approach is most likely to succeed or is least likely to fail. Thus I do not turn to Hobbes and Aristotle because their canonical status in the history of Western political thought leads me to conclude that they offer explicit answers that contemporary thinkers have failed to consider. In many ways, contemporary beliefs about the nature-politics relationship reflect commonplace understandings of how past theorists, including Hobbes and Aristotle, treated this relationship. Even where modern authors fail to acknowledge it, the superior depth and richness in the works of these earlier theorists allows me to explore the interactions between the conceptions of nature and politics that they employ.

Of course, there are potential dangers involved in a study of Hobbes and Aristotle driven by an interest in contemporary concerns. While ecological degradation has been a problem in many past societies, including ancient Greece and early modern Europe, contemporary ecological concerns were not apparent when Aristotle or Hobbes was writing. My intent is not to wrench these theorists from the context in which they lived and wrote in order to extract a direct commentary upon our contemporary dilemmas. This would stretch a reader's credulity to the breaking point. It is hard to imagine much more than inane platitudes or absurd simplifications surviving such an effort. Instead, my intention is to explore a relationship that was central to the philosophers in question—the relationship between the conception of nature they espoused and their own arguments about political life and political organization. The conceptions of nature with which they begin are certainly distinct from contemporary ecological ones and, particularly in the case of Hobbes, there are often stark contrasts between them.

While these differences are significant, it is the relationship between nature and politics that is key to all the theories I examine. It is this commonality that allows the sort of comparative study that I undertake. The widely differing contexts and conceptions of nature and politics found in the works considered here become a strength, rather than a weakness, of my study. By a comparative study of this conceptual relationship, the opportunity for more general conclusions and wider insights becomes greater. To the extent that I find convincing evidence of a dialectical relationship in these various works, this offers relatively firm grounds (though certainly not a proof) for believing that this sort of relationship is inescapable, rather than just the product of a few, perhaps poorly developed, contemporary arguments.

Situating My Analysis

Here I wish to clarify the relationship of my analysis to several other positions and debates. First, there is a sense in which my critique of an appeal to nature, and of the inattentiveness to conceptions of politics, may seem overstated as a characterization of contemporary ideas. In fact, some significant works by environmentalist thinkers in the 1990s, to which I alluded at the outset, have been highly critical of what they take to be the apolitical character of much theorizing in this field.[23] Moreover, it could surely be argued that even the most misguided interpreter of Hobbes or Aristotle could not fail to recognize the preeminent interest in politics exhibited by these philosophers. In each instance, this is undeniably the case.

Yet, while appealing to "political theory" or criticizing the apolitical character of past theorizing is valuable, it is not sufficient. With reference to the new environmentalist works in particular, I seek to build upon and extend their insights by explicating how deeply rooted this failure to take politics seriously can be. The problem can be more troublesome than even those who highlight political phenomena make clear.

A second question to address here focuses on the connection between theory and practice. While the critique that I develop centers on a failing of theory, my interest in developing it emerges from a concern with practice. This might raise doubt in some readers' minds as to whether the oft-heard refrain that "while it may be true in theory, it is not in practice" might rebut my critique. However, my critical analysis is rooted in the argument that neither a derivation of politics from nature nor a dualism that separates the two works *even* in theory. Conversely, the presumption behind the refrain is that while theory may be orderly, coherent, and subject to logical analysis, practice is far less likely to be so. No doubt. The result of pointing to an important and often unrecognized element of ambiguity within theory, however, is to bring it *closer* to practice. An emphasis on the messiness of the practical realization of abstract or theoretical goals would surprise no one. Nor would it seriously challenge a contention that a particular (derivative or dualist) relationship *could* or *should* exist, at least in theory. On the other hand, by arguing that such a relationship cannot be maintained in even some of the most renowned theoretical efforts that seem to many to do so, there is good reason to conclude that such a relationship also cannot exist in practice.

A third concern is the so-called "is/ought problem." I do not intend to focus in any detail on this problem or on the supposed "naturalistic fallacy" that has absorbed the efforts of so many moral philosophers since David Hume first raised the former, and particularly since G.E. Moore advanced his argument against the latter. If we seek to relate a conception of nature to a political order, it is because we have already concluded that that conception of nature ought to be adopted.

How can such a conclusion be reached, though? Those preoccupied with the is/ought problem typically find themselves compelled to make this question central to their investigations.[24] Does it require a belief in the existence of objective, intrinsic value in nature? Can it be reached via a subjective attribution of intrinsic value? Is a commitment to human self-interest, and thus an instrumental valuation of nature, sufficient? Is the latter truly a *moral* position? These questions dominate the debate among many environmental ethicists and, in more general form, play a role in many interpretations of Hobbes's and Aristotle's efforts to relate nature and politics. As a result, there is a need to touch upon them from time to time. Nevertheless, all the subjects of my study *do* embrace some conception of nature as meaningful. By adopting an essentially pragmatic perspective as to why or how they do so, I am able to set aside these seemingly intractable questions in order to shift attention to what I take to be more important and interesting and, it is hoped, more tractable ones.[25]

Finally, I wish to situate my arguments in relation to some significant strands of postmodernist discourse. As discussed, one of the key conceptual terms in my study is "nature." This is a central concept within the political and environmentalist thought that I examine, even when its meaning is not clearly fixed. Much written under the rubric of postmodernism has been acutely sensitive to the dangers of reifying "nature." As one author explains,

The history of ideas about nature, and the relationship of human cultures to nature, is so extensive that it would be naïve to think that any political tendency could claim ecology for its own.[26]

As a consequence, postmodernists seek to "throw the idea of nature and its alleged lessons into doubt more thoroughly and relentlessly than . . . ever."[27] My argument is clearly sympathetic to this project. The question, however, is what we do, or where we go, once we arrive at this conclusion. The often playful and occasionally mean-spirited critiques advanced within postmodernist writings have rarely addressed this question in a sustained or constructive manner.[28] Our goal must be to struggle with the meaning that nature should continue to have for us, even after we reject a directive role for it.[29]

It is when we encounter the constitutive role, I believe, that we should restrain our questioning of ideas of nature. The notion that in some form nature does constitute us and our communities is one that ought not be thrown out in a critique of foundational claims. Again, it is important to emphasize that the recognition of nature as constitutive need not and should not be collapsed into any sort of argument on behalf of environmental determinism. It is just such a nondeterministic, constitutive understanding that ecological thought at its best can offer us. With this understanding, we can reconstruct the relationship between nature and politics. Once we have done this, we can also advance more compelling political arguments about the resolution of problems that emerge from our understanding of nature.

Despite its insight, therefore, the contribution offered by the postmodernist critique of nature is often limited in two related senses.[30] First, the critique is limited by its generality. By failing to distinguish effectively between constitutive and directive roles for nature, it may end up "throwing out the baby with the bath water." By doing so, it also precludes any opportunity to learn from earlier attempts to relate nature and politics. Second, by focusing so singularly on a critique, it may neglect consideration of a more constructive effort to relate nature as constitutive to political argument.[31] In the chapters that follow, I consciously straddle the two extremes of dismissing nature as a subject of concern and promoting it as a directive principle for politics and society. It is in the middle ground that I believe we have the greatest potential for improving our understanding and hence our ability to engage in appropriate action.

Outline of the Book

At first glance, the order of my comparisons may seem peculiar or at least unchronological; a scan of chapter titles shows that I treat Hobbes's philosophy before Aristotle's. This work begins, however, in the present. It is contemporary environmentalist thinking that prompted my exploration of the history of Western thought. As such, my studies of Hobbes and Aristotle can be seen as critical and yet reflective of the contemporary search for the roots of environmental problems. Seen in this light, I hope that the movement from contemporary thought back to Hobbes and then further back to Aristotle will appear more sensible.

In chapter 2, I take a first look at the approaches of some classic and relatively well-known environmentalist thinkers. I focus on the arguments of Aldo Leopold, Lynn White, Robyn Eckersley, and Murray Bookchin. Although they differ in many other respects, I illustrate the ways that they come to regard their own visions of a future social and political order as deriving from a particular conception of nature. For these thinkers, nature—viewed through the lens of ecology—should be at the center of a worldview from which other arrangements can then be derived. By approaching the matter in this way, they neglect the independent influence of one's conception of politics on arguments for organizing human affairs. Of particular importance, I argue that even theorists who are outspoken in their criticisms of the apolitical character of other environmentalist writings, most notably Murray Bookchin, are guilty of regarding politics as derivative.

While chapter 2 focuses on arguments about the future, these are integrally related to our understanding of the present and the past. Chapter 3 considers dualist and derivative interpretations of the nature–politics relationship in the history of Western thought as they are found within environmentalist writings. While nonhuman nature has been frequently devalued in Western thought, I reject the claim that a radical separation or dualism has been responsible for this. By contrast, those who advance the derivative claim suggest that the failing of past thought lies not in a dualism, but in the mistaken conception of nature from which political arguments were derived. Here we are offered a basis for considering the nature–politics relationship in its various incarnations throughout Western thought. Nonetheless, I find the derivative characterization of this relationship problematic as well.

My focus in chapter 4 is on the work of Thomas Hobbes. Hobbes's political philosophy is far more widely known today than his works of natural philosophy. Nonetheless, the mechanistic conception of nature that he explicates in the latter plays a key role in his claims for the justification and distinctiveness of the former. In Hobbes's work, we find not only a supremely well-developed political philosophy but also a self-conscious assertion that this was derived from the new conception of nature that emerged during this era, and that Hobbes himself devoted much of his very long life to developing and promoting.

Despite his assertions, a number of interpreters have concluded that Hobbes was after all a dualist. The quest for an adequate understanding of the nature–politics relationship in his work obliges me to take account of these contrary arguments as well. Doing so leads to a closer consideration of the meanings that Hobbes attributed to nature and the relationship of this "nature" to his political argument. I argue that Hobbes does not successfully derive the latter from the former, yet he also was not a dualist.

This argument prompts us to distinguish between Hobbes's deterministic, mechanistic conception of nature itself and his conception of the "state of nature" as a condition to be transcended through the establishment of political order. The "state of nature" does, in fact, serve as a basis for Hobbesian politics by identifying those principles to which an adequate political order must adhere in order to transcend it. The "state of nature" is not, however, distinctly natural in Hobbes's sense of this term. In fact, I argue, the "state of nature" is a concept that can only be defined in relation to his prior conception of politics as the "power able to over-awe them all."[32] If this is so, then nature is unable to play the role of standard or authority for politics. Nonetheless, I argue, this does not deny a role for Hobbes's mechanistic conception of nature. The role is ultimately constitutive, however, not directive within his theory.

Chapter 5 turns from Hobbes to the source of the conception of nature that he and other participants in the scientific revolution rejected. Aristotle's articulation of a teleological nature stands in marked contrast to the mechanistic ideas that emerged in the seventeenth century. Indeed, in some respects Aristotle's conception of nature is quite amenable to contemporary environmentalist concerns and critiques.[33] If, as he is often interpreted, Aristotle suggests that nature and natural teleology can succeed in offering us a definitive standard for the construction of the best political regime, then he would have succeeded in establishing his conception of nature as the sort of directive principle from which politics can be derived. Yet Aristotle cannot make such a claim convincingly. Indeed, whether he does make this claim is open to question: Dualist interpreters contend that Aristotelian politics is distinguished by its antinaturalism.

Drawing insights from both of these lines of interpretation, I conclude that Aristotle can identify the *polis* as natural because it is viewed as the natural end of humans to be a part of this sort of social formation. As such, nature is constitutive of the *polis*. Yet to arrive at any conception of a best regime, Aristotle must rely upon a view of politics that establishes whether or not the latter is inclusive of what we would term economic and social realms. Of course, Aristotle does hold such a view. It is important that I show that this view is not a reflection of his political naturalism.

The goal of the sixth chapter is synthetic and reconstructive. Having established that an appropriate understanding of the nature–politics relationship can be reduced to neither a dualism nor derivation, I draw together and develop the more positive elements of my analysis suggested in previous chapters. The intellectual insight that drives arguments on behalf of both dualism and derivation does have some merit and this can best be appreciated by examining the way in which our conceptions of nature and politics interact with each other. Here I develop the significance of one's experience and relationship to place—where we live, work, and play—as both a product of this interaction and a key influence on our understanding of what is politically necessary and appropriate.

All of this invites questions about how environmental politics might be reconceived in light of the position developed to this point. In my seventh and final chapter, I illustrate how struggles as diverse as those over toxic waste dumps in poor and minority neighborhoods, rainforest protection in Brazil, and land use in the American West point us toward a broader conception of environmental politics, once we escape the sway of derivative and dualist interpretations. Movements such as these build upon the experiences and places from which they emerge. They, like more familiar environmentalist efforts to protect wilderness or endangered species, are inherently political movements shaped by our embeddedness in nature. As a result, they help illustrate the value of reconstructing the relationship between nature and politics in the manner that I pursue throughout this book.

I

Future and Past

2

Worldviews and the Evasion of Politics in Environmentalist Thought

The mainstream political view of "the environment," at least in contemporary Western liberal-democratic societies, is that it is an issue area. This "issue area" is understood to include a number of particular concerns, including toxic pollution, forest destruction, biodiversity loss, global warming, and so forth. Viewed in this manner, environmentalists are seen as representing a particular interest—one among many—that any nominally democratic or pluralistic political system should consider when making policy. The consequence of this mainstream view has been that while environmental concerns are now a recognizable part of the political landscape in a great many places, they often have been politically marginalized by powerful economic, social, or national security interests.

Much of the philosophical work on environmental concerns reacts against the limitations of this view of the environment as one issue area among many in a pluralist system. Despite important differences among and between deep ecologists, ecofeminists, social ecologists, environmental ethicists, and others, these critical environmentalists share the conviction that this contemporary mainstream view must be exposed as inadequate if we are to achieve a more ecologically sound future. I share a frustration with the limitations of "issue area environmentalism" and I also share the conviction that challenging these limitations is key to environmentalist political thought. Where my analysis diverges from many schools of environmentalist thought, however, is at the point where they assert or presume that their challenge must be based upon an alternative worldview.

Dichotomous Worldviews and Natural Standards

Whether the dichotomy is between "shallow" and "deep ecology"[1]; "environmentalism" and "social ecology"[2]; "light green" and "dark Green"[3]; or "anthropocentrism" and "ecocentrism,"[4] the boundary delineated by environmentalist thinkers is between this familiar brand of issue-area environmentalism and an alternative "environmentalism" (for lack of a more agreed-upon term) understood as the basis for a new and encompassing worldview.[5] The consequence is that the transformation of the existing, dominant, and encompassing worldview becomes something of a holy grail for many critical environmentalists.

Dichotomous worldviews are used to explain both our present environmental dilemmas and future resolutions of them. It is a new conception of nature that becomes the basis here for the new worldview.[6] A worldview itself is understood to express a distinctive attitude and hence a distinctive relationship between humanity and the rest of nature. Crucially, a worldview is regarded as capable of transforming human politics and society. Nonetheless, many environmentalist writers are sensitive to the criticism that they either have committed a "naturalistic fallacy" or that claims to derive politics from nature can provide justification for such evils as Social Darwinism or even Nazism.[7] As a result, they explicitly contend that one cannot derive normative conclusions about human society directly from nature or ecological insights. The former, they say, can only be "inspired by" or "grounded in" nature or ecological science.[8] While these qualifications are quite valid, the distinction between being "inspired by" nature and deriving a normative view from nature serves, on closer inspection, to undermine the initial claim to delineate a worldview uniquely able to address environmental crisis.

It is illogical to assert that only a particular worldview can be inspired by an ecological conception of nature. Inspiration does not work that way. An ecological conception of nature, like any other conception, has the potential to inspire a number of diverse and potentially contradictory positions. While there are a variety of grounds upon which we might criticize or reject these positions, we could not do so on the assertion that they have not truly been inspired by a particular conception of nature.

On the other hand, it is at least logically consistent (although I argue incorrect) to maintain that only one view can be correctly derived from such a conception. Only then does it make sense to assert that one worldview is consistent with an ecological conception of nature, while all contrasting views are inconsistent with it.

Where an ecological conception of nature offers the basis for the new worldview that environmentalists hope will dominate the future, the absence of this conception is understood to characterize the worldview of the present. A dichotomy is thus established between the future ecological worldview and the present nonecological worldview. My goal in the remainder of this chapter is twofold: to show the centrality of this dichotomy in environmentalist writings and to draw upon each author's insights to illustrate the inability to sustain this argument.

Both Aldo Leopold and Lynn White formulate characteristic and influential positions that reveal the core of this response. Thus, a consideration of their arguments can offer insight into a way of thinking about environmental concerns that resonates widely among environmentalist philosophers.

As I have already noted, a number of more self-consciously political theorists have criticized fellow environmentalists on grounds similar to those that I describe here. One of the best known and most unforgiving of these critics has been Murray Bookchin.[9] Robyn Eckersley offers a similar critique that includes Bookchin within its scope.[10] Following my consideration of Leopold and White, then, I examine the arguments of both Eckersley and Bookchin. In doing so, I highlight their continued reliance upon nature as a standard from which to derive ecopolitical arguments, despite their explicit criticism of others for doing so. That even such attentive thinkers could embrace this strategy ought to suggest the enormous and sometimes subliminal allure it can hold for environmentalists.

Classics of Environmentalist Thought

Aldo Leopold's *A Sand County Almanac* has become "the intellectual touchstone for the most far-reaching environmental movement in American History," according to environmental historian Roderick Nash; "a famous, almost holy book" in the words of novelist Wallace Stegner.[11] The final essay, "The Land Ethic," has been particularly influential in the development of philosophical thought on environmental concerns. J. Baird Callicott has described this essay as:

the first self-conscious, sustained, and systematic attempt in modern Western literature to develop an ethical theory which would include the whole of terrestrial nature and terrestrial nature as a whole within the purview of morals.[12]

While no one who reads the book can fail to be struck by the power of Leopold's personal and poetic reflections upon the land, his most distinctive contribution was to argue that the new science of ecology—of which, as a professor of ecology and chair of game management, he was a leading exponent—offered a firm foundation for an ethical concern for "the land."[13] As Stegner observed, "it was not for its novelty that people responded to Leopold's call for a land ethic. It was for his assurance . . . that science corroborates our concern."[14]

Leopold's analysis is a response to a paradox. Concurrent with the "growth in knowledge of land" and "good intentions toward land" has been the expanded "abuse of land."[15] The attempt to understand the causes of this paradox led Leopold to focus on a dichotomy between two distinct worldviews. In an uncommonly prosaic moment, Leopold labels the dichotomy the "A-B Cleavage."[16] While "As" regard nature merely as the basis for commodity production, "Bs" regard its value more broadly and exhibit "the stirrings of an ecological conscience."[17] Leopold explains that

In all of these cleavages, we see repeated the same basic paradoxes: man the conqueror *versus* man the biotic citizen; science the sharpener of his sword *versus* science the searchlight on his universe; land the slave and servant *versus* land the collective organism.[18]

Inherent in Leopold's argument is the conviction that a worldview characterized by a particular relationship between humans and the land is defined by one's conception of nature itself. One view holds that nature is merely a collection of discrete organisms, and consequently that these are valuable only in a narrowly instrumental and economistic manner. The other—based on the understanding of ecological relationships that Leopold develops throughout his book—concludes that there is greater noneconomic and noninstrumental value in the land.

Leopold argues for expanding the "moral community" within which humans are located to include the land. He perceives the expansion of communal boundaries to be the culmination of an evolutionary process. Over time, Leopold contends, philosophers and others came to understand the moral community as encompassing the whole of the human species. The great contribution of ecological science, Leopold believes, will lead to a further expansion of the moral community to include the land. As a result, he argues that we will:

change . . . the role of *Homo sapiens* from conqueror of the land-community to plain member and citizen of it. It implies respect for his fellow-members, and also respect for the community as such.[19]

Central to this understanding is the belief that a broadening of boundaries of our moral community will redefine our relationship with those entities newly defined as community members. For Leopold, particular ethical, social, and political relationships seem to flow unproblematically from the definition of communal boundaries. While the "A" worldview is commensurate with a boundary line drawn tightly around the human community, the expansion of this boundary to include the land will be synonymous with the realization of the "B" worldview that Leopold describes.

This is reflected in his casual yet distinctive use of the term "citizen" in the above quotation. "Citizens," members of a common community, in Leopold's usage, do not seek to act as conquerors toward each other. He devotes little attention to the possibility for conflict or disagreement within the boundaries of a community. Instead, Leopold suggests that the adoption of a new conception of nature will provide us with the principle upon which our worldview will be reformulated, and offers little reason to believe that he anticipated that significant problems would remain if this were to be achieved. Why might that be?

To be convincing, Leopold's position must assume that our worldview, guided by our conception of nature, is a unified whole that can direct our social and political actions in a seamless and consistent manner. Only if this is true can Leopold appropriately neglect the possibility of conflict or disagreement within an accepted conception of community. Yet this assumption cannot be based on the ecological insights offered in his writings.

Indeed, it is a mistake to assume that redefining the boundaries of our community would transform our practices toward the rest of nature, any more than the incorporation of humanity within such a moral conception alone has served to eliminate disagreement, conflict, or oppression among humans. He fails to consider the disparate perspectives and interpretations that draw upon different experiences within a common communal heritage or worldview. The choice among these is clearly a political one that cannot be resolved by a redefinition of communal boundaries alone.

Lynn White's essay on "The Historical Roots of Our Ecologic Crisis" offers a second especially prominent environmentalist view. As Peter Coates observed, "[t]he sheer weight of citations in environmentalist literature suggests that White's thesis is the most influential of all diagnoses."[20] Many have disagreed with the specifics of White's analysis, of course, but Langdon Winner is quite correct to argue that White provided "an invitation to a style of inquiry that a great many have taken to heart."[21]

White contends that although we live in a "post-Christian age," the substance of our thought, and the development of our technology and society, remains reflective of Christian doctrine.[22] He describes the victory of this doctrine over paganism as the "greatest psychic revolution in the history of our culture."[23] As with Leopold, White suggests that our ecological problems reflect a particular worldview, and that a solution requires the same sort of "psychic revolution" that brought about this view in the first place. The superficial focus here is on religion, not nature, as the basis for our worldview. Yet it is not a change in religious conviction that White sees as sparking our next psychic revolution. Instead, it is a new understanding of nature, reflected in our recognition of ecological destruction, that challenges our present worldview. Here, White measures the appropriateness and even the truth of religious doctrine against the yardstick of "our ecologic crisis."

Much of the debate that followed publication of White's thesis in the late 1960s centered on whether Christianity was truly as anthropocentric as he seemed to assert, and to what degree other religious or philosophical traditions might be more commensurate with an ecological society.[24] This response makes sense because of White's apparent suggestion that the "historical root" and the resolution of environmental crisis can each be located in a particular worldview. That he characterizes such worldviews as religious is rather incidental to this more fundamental presupposition. To ask whether such a worldview is consistent with a concern for nature is in effect to maintain that this concern can provide us with a principle from which a comprehensive set of ideas, institutions, and practices will ultimately follow.

White's skill as a historian prevented him from consistently developing the position often attributed to him—that Christianity itself delineates a worldview destructive of nature. In fact, a close reading of his brief essay suggests that he views "voluntarism" and "democracy" (as he understands these terms) as ultimately responsible for ecological crisis, and understands these as rooted in a particular (Western) interpretation of Christianity. White makes clear that other interpretations or paths were possible within Christian societies.[25] Thus, White's analysis ultimately suggests a "worldview" that is far more diverse and malleable than it appeared at first. Certainly it is more malleable than much of the debate surrounding White's thesis would lead one to believe. Indeed, in a subsequent essay, White maintains that "every complex religious tradition [has] recessive genes which in new circumstances may become dominant genes."[26]

There is a significant tension between White's best-known claim that Christianity is the root of ecological crisis, and his more subtle position that "complex" religious traditions necessarily include diverse viewpoints. White's latter position points to social or political interpretation as more important than Christianity per se in setting the stage for environmental destruction. After all, if both dominant and recessive interpretations of nature are imbued within Christianity, then the most interesting investigation would focus on how this process of interpretation takes place, and why a particular interpretation gains dominance.

Environmentalist Political Theory

The recent blossoming of literature by "green" political theorists provides an important and refreshing contrast to much environmentalist thought. These works aim to take politics seriously and a number offer in-depth treatments of themes that include pluralism, social justice, rationality, virtue, and property rights.[27] Such writers would not appear to be susceptible to the analysis and criticism that I have developed so far. However, the matter is not simple. A close examination of two prominent theorists leads me to conclude that they, too, perceive nature as offering a basis for the transformation of our social and political order. Despite differences in emphasis, Eckersley and Bookchin are wedded to the same assumptions and arguments described in my discussion of Leopold and White.

Consider Robyn Eckersley's *Environmentalism and Political Theory: Toward an Ecocentric Approach*. As the subtitle suggests, the most central concept in this evaluation of political theory is "ecocentrism." She posits that:

> an ecocentric approach regards the question of our proper place in the rest of nature as logically prior to the question of what are the most appropriate social and political arrangements for human communities. That is, the determination of social and political questions must proceed from, or at least be consistent with, an adequate determination of this more fundamental question.[28]

Ecocentrism can constitute a worldview in the manner I have used the term here because it is meant to clearly specify "our proper place in the rest of nature." But where does our understanding of this relationship itself come from? According to Eckersley, it is rooted in a conception of nature as:

> *internal relatedness,* according to which all organisms are not simply interrelated with their environment but also *constituted* by those very interrelationships . . . This model of reality undermines anthropocentrism. . . . In this respect, ecocentric theorists, far from being anti-science, often enlist science to help undermine deeply ingrained anthropocentric assumptions that have found their way into many branches of the social sciences and humanities, including modern political theory.[29]

Thus our model of reality drawn from the insights of ecological science can serve as the basis for an ecocentric worldview. In turn, ecocentrism provides Eckersley with the standard by which to evaluate existing political theories.

It is worth noting that Eckersley also asserts that "appealing to the authority of nature (as known by ecology) is no substitute for ethical argument," thus disavowing any simple derivation of politics from nature.[30] In the context of her other statements quoted here, however, what can this disavowal mean? One clear meaning is a rejection of a deterministic relationship between ideas and social practices. She argues convincingly that there is "nothing *inevitable* about the possibility of a new, ecologically informed cultural transformation."[31] A second intent is to put distance between her work and the claim that scientific understanding of nature is itself a sufficient foundation for political theory. In her analysis, the relationship between a conception of nature and political argument is mediated by the account of ecocentrism. The entire structure of the book makes it clear, however, that politics should be derived from an ecocentric commitment.

While Eckersley provides insightful descriptions of the political theories that she considers, her framework prevents any real dialogue with them, as she intends. By using an ecocentric measuring rod to evaluate political theory, Eckersley wraps up most sections in her book with the conclusion that, "From an ecocentric perspective, however, . . . [this approach] simply does not go far enough: it is fundamentally limited by its anthropocentrism."[32] By focusing on the claim that existing political theories do "not go far enough," we might be led to the conclusion that once we get this far, an ecocentric approach will do more. In this case, distinctive ecocentric arguments would only be relevant once we have gone as far as existing, nonecocentric, approaches can take us.

This emphasis on compatibility and convergence, however, cannot be Eckersley's true intent. Instead, by asserting that existing approaches are "fundamentally limited by [their] anthropocentrism," she implies that a political theory not so limited would look fundamentally different. One drawn from an ecocentric perspective, we are led to believe, would be both different from all existing political theories, limited as they are by anthropocentrism, and distinctively the product of an ecocentric worldview. Here the emphasis is on incompatibility and divergence.[33] Rejection of other political theories from an ecocentric perspective really requires that Eckersley argue that they represent the wrong path to follow. This is obligatory, especially because the "emancipatory" political theories that she examines often regard *each other* as the wrong path, although for reasons that differ from Eckersley's.[34] It is really only if this claim can be made convincingly that the project that Eckersley sets for herself in this book can be justified.

Eckersley, however, does not put forward such a political theory. In the final two pages of her book she does offer a very brief "broad-brush picture of . . . an ecocentric polity" that includes a "multilevelled decision-making structure" combined with "greater dispersal of political and economic power" and an "ecocentric emancipatory culture."[35] Since her intent was to identify a distinctly ecocentric political theory, however, this sketch is a disappointment.

The only element in this political vision that distinguishes it from other "anthropocentric" political theories is her emphasis on an ecocentric emancipatory culture. Yet this cultural element is unsatisfying, since the original intent was to identify the "social and political arrangements for human communities"[36] that would flow *from* this cultural perspective. As a result, her critique of other theories is weakened. Eckersley's approach to politics and political theory assumes that it is possible to derive these from a commitment to ecocentrism. That she does not succeed is a reflection of something more than a hurried ending to her book. Instead, it reveals a flaw in the premise with which she began. Neither an ecocentric approach nor any other worldview can itself successfully delineate particular social and political ideas and arrangements.

A very different example of a politically oriented approach to environmentalist philosophy can be found in the lifework of Murray Bookchin. While Bookchin has advanced a scathing critique of deep ecology and ecocentrism on the grounds that it is dangerously apolitical,[37] he insists that this is the distinctive product of what he has labeled "social ecology." Although his writings are voluminous, the central thesis can be summarized quickly. It is that "the very notion of the domination of nature by man stems from the very real domination of human by human."[38] For Bookchin, ecological crisis offers a powerful incentive to pursue the transformation of social and political institutions and processes in a non-hierarchical, anarchistic direction. He maintains that this vision is not merely "wishful thinking." Either we will transform societal relations in order to achieve a "harmonized world with an ecological sensibility based on a rich commitment to community, mutual aid, and new technologies," or we will move toward an "apocalypse."[39] Thus, "what *should* be could become what *must* be, if humanity and the biocomplexity on which it rests were to survive."[40]

A central question for Bookchin's analysis is how it is that he can establish "what should be" in the first place. He has been highly critical of those whom he views as importing ethical standards from outside the context of their ecological analyses,[41] and has insisted that his own social and political vision can be and has been derived from a true understanding of nature itself. This argument is notable for two reasons. First, he is significantly more explicit and intentional than Eckersley and others in arguing that a true understanding of nature itself can provide the matrix within which to develop an objective ecological ethic.[42] From this, he argues that the social and political vision that he advances is uniquely natural or ecological. "Ecological" is a term of distinction for Bookchin, one that he applies only to approaches congruent with his own "social ecology." Second, he describes this approach as distinctively based on an "organic," "dialectical," and ultimately "ecological" form of reason.[43] This is meant to suggest that he may be immune to the charge of seeking to derive his social and political views from nature—a charge invited by the first claim.

Bookchin places heavy emphasis on his use of dialectical argument and the importance of social and political institutions and hierarchy in shaping ecologically destructive behavior. Given this emphasis, however, his most fundamental claims seem rather one-dimensional. He argues that a distinctive form of reasoning ("dialectical naturalism") is uniquely able to get to the truth about nature. It is this truth, he contends, that allows him to identify the "logic of differentiation" as a fundamental ecological requirement that is the same for both (first) nature and society.[44] Just as diversity, richness, and complexity can be understood as the *telos* of development in natural ecosystems, so should they be understood as the end of human institutions and modes of organization.[45] It is precisely this logic of differentiation, Bookchin contends, that has led to the evolutionary emergence of self-conscious, human, "second" nature from the mute, amoral world of "first nature."[46] Humanity and human reason, in this vision, are understood as "nature rendered self-conscious."[47]

For Bookchin, then, human action, second nature, can be construed as natural.[48] Thus an ecological ethic must advocate self-conscious human stewardship of the planet's first nature. It must deny either the possibility or the appeal of returning humanity to first nature. Were it possible, such a return would require us to deny the conscious and reflective qualities that distinguish us from it.[49]

Bookchin is eager to argue that the character of second nature currently is destructive—a moral judgment that emerges from this nature—yet argues that the existence of such moral judgments means that human destructiveness need not be prevalent in all social orders. His argument also relies upon an interpretation of history in which early "organic societies" are viewed as having lived an essentially nonhierarchical, egalitarian, cooperative existence that was subverted with the rise of hierarchical institutions and ways of thinking.[50] Bookchin envisions a world in which second nature has been transcended into "free nature" in an "ecological society."[51] Here, humanity has overcome the false dichotomy between the realms of nature and freedom that was introduced with the rise of hierarchy and civilization. Instead, freedom now is found within the realm of nature and necessity, rather than by overcoming and dominating it.[52]

The logic of differentiation does more for Bookchin than justifying human stewardship. It is also meant to provide the foundation and justification for his vision of a nonhierarchical anarchist society. As such, it is presented as nature's principle, uniquely discoverable through a process of ecological reasoning: "dialectical naturalism." Bookchin contends that overcoming human domination and hierarchical institutions is necessary for reconciling humanity with the natural order itself. While he is quite aware that a wide variety of moral, social, and political "imperatives" claim to be drawn from an interpretation of nature, he consistently maintains that it is only "nature conceived by a radical social ecology" that has truly understood the meaning inherent in natural development.[53]

According to Bookchin, once a free and nonhierarchical society is established, the foundation for an ideology of dominating nature will have been removed. By focusing on this one important element in Bookchin's argument, it seems easy to conclude (as he asserts) that he is reversing the arrows of other environmentalist arguments, and thus placing much greater emphasis on social and political institutions. However, as we have already seen, the justification for the free society that Bookchin envisions is itself found in his understanding of ecology. Thus, a more complete picture of his argument would begin, not with his vision of social and political freedom, but with his conception of ecology, which offers him a distinctive basis for defining and defending the societal vision.

Yet once this point is acknowledged, Bookchin's argument appears to be circular. A new (ecological) conception of nature (1), leads to a new, nonhierarchical, free society (2), which in turn leads to a transformation of our ideology of dominating nature (3). But what would this transformation lead to? A new, ecological, conception of nature. Thus, the first and third elements in his argument appear to be the same. In fact, what he seems to want to suggest is that a nonhierarchical, free society (2) would lead us to transform our *acts* of domination toward nature (3). For good reason, however, he does not make this argument explicitly.

A distinction between the act of dominating or destroying nature and the ideology of dominating nature can be found in many of Bookchin's works. The distinction is often made in an effort to rebut the accusation that he is making a deterministic argument.[54] Convincing in its own right, this distinction leads to the conclusion that the elimination of an ideology of domination, overwhelmingly difficult though it may be, cannot itself alter our actions, and that the existence of egalitarian institutions does not ensure the adoption of ecologically sensitive action—both points that Bookchin acknowledges.

It is interesting that he concedes that undermining the ideology of domination of nature, while necessary, may not be sufficient to end ecological exploitation. This is because what he terms "transient behavior patterns" have often led to ecological exploitation even where this ideology of domination was absent. Yet if this is the case, then the preoccupation with eliminating this ideology seems misplaced. It seems likely instead that the role played by social and political ideas and institutions can be more variable than Bookchin expects, and thus important in ways that he fails to anticipate.

Bookchin's ultimately limited appreciation of the role played by politics and political interpretation is of course a familiar problem by this point. The failing is especially notable in his case, however, because of his own awareness and criticism of many key elements of it. That even he fails to incorporate an adequate role for interpretations and judgments about politics within his theory ought to suggest to us the great difficulty in doing so.

Conclusion

The perspective commonly adopted by environmentalist philosophers is now I trust a bit clearer. Because the notion of encompassing yet dichotomous worldviews is so central to this approach, political debate becomes derivative and hence largely inconsequential. This is the case both within existing societies and within the ecological societies envisioned by these theorists. Yet this framework is unable to contain or prevent, even in theory, the interpretive differences and diverse possibilities that only political judgments can resolve.

We live in a world with many cultural and moral traditions, and more than one conception of, and attitude toward, nature. As I have emphasized, environmentalist thinkers often must acknowledge this, but then they fail to account for the implications of this diversity in their overall argument. As a result, they seriously overestimate the importance of transforming our worldview—a transformation that appears dependent upon the adoption of a new conception of nature.

Overlooked or underestimated are the social and political processes that result in certain attitudes toward nature becoming prominent at a given place and time. Many views and attitudes go unrecognized, or at least do not become dominant, because they aren't made visible by existing political processes and decisions. If so, then change does not require a paradigmatic shift as much as a political one.

3

Searching for Roots: Environmentalist Interpretations of the History of Western Thought

The environmentalist position described in chapter 2, which favors the adoption of a new ecological worldview, is ultimately dependent upon the reinforcement that it receives from two influential interpretations of the past. The past here is the history of Western thought, since it is Western institutions and societies that are typically regarded as having most profoundly shaped the current relationship between humans and nonhuman nature. These two interpretations—which I term the *dualist* and the *derivative*—are highly seductive to environmental theorists, yet they are mutually exclusive and problematic. Their seductiveness lies largely in the role each plays in encouraging a particular prescription for the future. Once we recognize the difficulties with these interpretations of the past, the faults in their prescription for the future will also become more apparent.

The insight at the core of diverse characterizations of a new ecological worldview is one drawn from ecological science—that humans and nonhuman nature are necessarily connected and hence interdependent. Indeed, this emphasis on interconnectedness often appears to be a truism, a claim seemingly impossible to deny if it is subjected to a moderate degree of reflection. And yet environmentalist theorists identify precisely this denial as a root cause of contemporary environmental crises. How can this seeming inconsistency be explained? Two answers are suggested.

One is that Western thought, up to the present, is fundamentally antinatural, hence *dualistic*. A firewall exists in our thinking; one that isolates human culture and politics from the rest of nature or the environment. While this dualism is said to be mistaken, it is characterized as such a deeply rooted element in Western thinking that it obscures even obvious truths such as this. Seen from this viewpoint, an acceptance of interdependence would oblige us first to overturn the entire historical legacy of Western thought and then to transform our worldview.

A second answer is that Western social and political thought has been naturalistic all along, in the sense that "nature" has been the touchstone of legitimacy for social and political authority, but that the conceptions of nature from which this tradition of thought has been derived are fallacious or at least misguided. From this contrasting perspective, the fundamental problem has not been the dualism of past thought (in which nature was ignored in the consideration and construction of human societies), but instead the "unecological" character of past and present conceptions of nature. Here the parallel between an interpretation of the past and the envisioned ecological worldview is greater. Just as the acceptance of an ecological conception of nature is said to offer the basis for this worldview and a corresponding future society, the acceptance of either a mechanistic or an anthropocentric conception of nature is said to be the basis for our current worldview and hence for our society's destructive relationship to the natural world.

The dominant conception of nature is certainly regarded as problematic here in that it is said to result in the devaluation of the nonhuman world. Rather than arguing for the need to introduce a conception of nature into our thinking, however, this second approach asserts that some conception of nature always has been at the core of our worldview. The challenge in this case is to replace this conception with a suitably ecological one. Once again, the assumption is that appropriate social and political change will follow from this replacement. I characterize this naturalistic view of Western social and political thought as a *derivative* account, one in which some image of nature serves as the normative principle from which social and political order is derived.

It is important to emphasize that the dualist and derivative accounts are not offered as descriptions of distinct traditions in the history of Western thought. Instead, they are divergent characterizations of the same, inclusive Western history. It turns out that the treatment of Western thought as essentially unitary in this sense is central to the functional role that both the dualist and derivative accounts can play in contemporary environmentalist thought.

Discussing the past in terms of either worldview supports the suggestion that real change will result only from a transformation of this view.

This transformation appears necessary because the character of the current worldview seems to preclude meaningful change within its parameters. The transformation appears possible precisely because it is based on the seemingly quite powerful ecological insight of the inescapable interdependence between humans and nonhuman nature.

The inadequacy of viewing both the past and the future in terms of worldviews or paradigms becomes clear if we seek to understand the role of politics within such a framework. Politics becomes a mere consequence of our worldview. There is no space here for a discussion of either political ideas or structures as an influence upon the character of our relationship with the nonhuman world, a highly ironic yet logical consequence of a vision in which politics and society are to be transformed, yet are regarded as derivative of a worldview. There is much reason to believe that this influence is central and inescapable.

In chapter 2 we saw that even thoughtful and consciously political works, such as those by Eckersley and Bookchin, often evaded political judgments by overemphasizing the role of an ecological worldview. Other recent environmentalist political thought has made an even greater effort to distance itself from "worldview" arguments.[1] Valuable though they are, however, these analyses do not address—and sometimes fail to appreciate—the depth of support that this evasion of politics receives from the dominant interpretations of the past.

In the remainder of this chapter I proceed as follows: First I outline the dualist account of Western thought as articulated in environmentalist thought, and then examine why this account is often so seductive. Following this, I sketch the derivative account, followed by a similar examination in which I discuss the striking functional similarities of these otherwise divergent interpretations. Neither of these two accounts is fanciful, and support for each has been presented by many authors. Yet the accounts are mutually exclusive and I argue that ultimately neither is adequate. Western thinkers have often devalued nature, but neither dualism nor derivation has been a consistent reason for this devaluation. While a recognition of this problematizes the theoretical project needed to address the very real concerns of environmentalists, its more vital contribution is to allow us to move questions of power and political judgment to center stage in a more coherent and sustainable manner than they have been before.

Dualism

The dualist interpretation does not simply assert that Western thinkers have typically distinguished between humanity and nature, or that some prominent thinkers have characterized these two categories as opposed. Instead, I argue that this interpretation necessarily regards a deeply rooted duality between humans and nature as endemic in Western thought. As a consequence, Western social and political thought is regarded as humanistic—contained on the human side of this dualism— and thus completely removed from the world of nature. Regarded in this manner, current social and political thought appears completely unable to address the challenges and problems posed by our interaction with this nonhuman world.

Val Plumwood offers an especially developed account along these lines. Drawing upon feminist insights into the gendered character of divisions between "public" and "private," she argues that there is a commonality between these and all other dichotomies or dualisms to be found in Western thought. She seeks to trace the origins of dualistic thinking regarding human–nature relations to the very beginning of Western philosophy:

the Platonic, Aristotelian, Christian rationalist and Cartesian rationalist traditions [all] exhibit radical exclusion as well as other dualistic features . . . [this] facilitates the conclusion that there are two quite different sorts of substances or orders of being in the world. . . . *There is a total break or discontinuity between humans and nature, such that humans are completely different from everything else in nature.*[2]

It is the emphasis on "total . . . discontinuity" that it is especially important to notice. It is this total character of the distinction between humanity and nature that justifies Plumwood's characterization of it as a dualism. As a result, a view that could undermine such an all-encompassing dualism would also appear to transform Western thought.

In a similar vein, Peter Marshall asserts that "the traditional Western worldview" has been:

supported by the mainstream Judeo-Christian tradition . . . [and by] the rationalist tradition from Plato onwards [which] also separated the mind from the body, the observer from the observed, and humanity from nature.

By contrast, he maintains that:

A new vision of the world is emerging which recognizes the interrelatedness of all things and beings and which presents humanity as an integral part of the organic whole.[3]

In this view, it is because the cleavage between humanity and nature is coextensive with Western thought that a seemingly simple recognition of "the interrelatedness of all things and beings" becomes sufficient to pose a fundamental challenge to the tradition.

Others agree about the pervasive and problematic character of this dualism, yet focus on less ancient sources of its emergence. Here, "modernity"—as epitomized in the mind–body division of Descartes, or alternatively the emphasis on artifice inherent in social contract philosophy— is identified as the primary source of dualistic and hence antinaturalistic thinking.[4] Thus in a chapter of his book *Thinking About Nature* entitled "Beyond the Social Contract," Andrew Brennan can argue that in dramatic contrast to the contractarian view, a new ecological theory of value can allow for:

the recognition that all human life is lived within some natural context and that it is in terms of that context that the identities of very different human lives are forged.[5]

Brennan is surely right to contend that an appreciation of ecology promotes the "recognition that all human life is lived within some natural context," just as Marshall is to argue for a recognition of the "interrelatedness of all things and beings." This emphasis is only truly distinctive, however, if we also insist that other views, those not informed by ecology, *necessarily* ignore or deny any natural context. To do so is to adhere to an interpretation in which previous thinkers regarded the basis for human society as isolated from *any* conception of nature (not only one informed by the contemporary science of ecology)—a dualistic view here attributed to the social contract approach that Brennan seeks to move "beyond."

In sum, the characterization of Western thought as inherently dualistic is a familiar explanation among environmentalist thinkers for the devaluation and exploitation of nature in Western societies. The question of whether this dualism is the product of something distinctly modern within Western thought appears more ambiguous. Yet one might be forgiven for concluding that the central place of a human–nature dualism in at least the modern expressions of Western thought was a rather settled question.

The Allure of the Dualist Interpretation

An "ecological" view—one that speaks about interrelatedness and natural context—appears to be entirely novel in light of the dualist account because existing thought is seen as failing to acknowledge *any* role for nature in its reflections upon humanity and human society. The perceived novelty of the ecological view within the Western tradition suggests that even if it is only formulated in a vague or abstract manner, it can have great power to restructure our thinking on a wide variety of other subjects; most notably on politics and social order. At the same time, and at least in the absence of specific elaboration or conclusions, the simple assertion that humans are integrally related to the nonhuman natural world seems unlikely to provoke serious rebuttal. Indeed, it is a virtual truism. It can be elevated to play such a singularly important role in environmentalist thought precisely because—and only if—the Western tradition is understood to have previously denied or rejected its truth.

It is the unique combination of these two disparate and seemingly incompatible elements—novelty and incontrovertibility—that makes this account seem especially attractive. Because the view is regarded as novel, it seems to pose a revolutionary challenge to existing thought. Yet because it seems relatively uncontroversial and wholly incontrovertible to acknowledge the interconnectedness of human and nonhuman nature, this challenge would also seem to have tremendous potential for success. It allows one to point to the "apparently obvious truth" that we are a "part of nature" as a concept with revolutionary implications, if only it were truly accepted.[6] The "obviousness" of this truth makes it quite attractive as a basis for promoting social and political change. Who among us would wish to deny—as if we could!—that humans are a part of nature? If acceptance of this point is all that is necessary to produce a society that refrains from ecological devastation, then such an effort would appear quite likely to succeed.

And yet it is precisely the "obviousness" of the ecological view in question that demands an explanation for its apparently unwelcome reception, or at least its lack of widespread acceptance, to this point. The characterization of Western thought as essentially consisting of a human–nature dualism offers one powerful explanation. If this dualism is key, then the acceptance of an otherwise obvious truth about human embeddedness within nature is constrained by the full weight of our intellectual heritage. Any effort to transcend dualism would require that we first overturn the entire structure of Western thought. Once again, Plumwood makes this contention especially explicit by arguing that an array of distinctions to be found in Western thought are all connected to the same fundamental structural duality that permeates this thought. Neither oppression based on gender nor exploitation of nonhuman nature is offered as a singular source of Western culture's ills. Instead, Plumwood presents each as a particular manifestation of a broader and more encompassing dualism at the core of Western thought. She maintains that

key elements in the dualistic structure in western thought are the following sets of contrasting pairs:

culture/nature
reason/nature
male/female
mind/body (nature)
master/slave
reason/matter (physicality)
rationality/animality (nature)
reason/emotion (nature)
mind, spirit/nature
freedom/necessity (nature)
universal/particular
human/nature (non-human)
civilized/primitive (nature)
production/reproduction (nature)
public/private
subject/object
self/other[7]

It is because of this fundamentally "dualistic structure in western thought" that an argument on behalf of the interconnectedness of humanity with the rest of the natural world is interpreted as such a challenge. Moreover, the many manifestations of this structure that Plumwood outlines seem to suggest just how well fortified it is, and hence the great difficulty involved in overcoming dualism. This offers an explanation for the difficulty in advancing ecological concerns. If, on the other hand,

Plumwood is wrong, and dualism is not so pervasive in our thinking, then the explanation for the inadequacies of environmental action cannot be overcome by an appeal as "obvious" as the claim that humans are interrelated with the rest of nature.[8]

For the dualist interpretation of Western thought to serve the key explanatory role often assigned to it in environmentalist writing, it must be seen as overwhelmingly dominant in Western thought. Suggesting otherwise would intimate a more ambiguous and potentially contradictory record of Western thought in this regard. It is necessary to view dualism in this manner because of the connection between theory and practice implicit in an appeal to it as an explanation for our environmental ills. It is environmental *practices* that are ultimately the subject of concern, even if they are not the subject of analysis. For theory to serve as an explanation for these practices in the manner suggested, it must be something of a deep structure—as talk of a worldview or paradigm is meant to suggest—that molds our practices in important ways and prevents alternatives from being considered. If the theoretical view that is said to explain our environmental practices is a human–nature dualism, then existing theory must speak with one unequivocally dominant voice with regard to it. The acknowledgment of influential alternative voices would pose a threat to the clarity and coherence of this argument.

If dualism were not clearly dominant in Western thought, then it could not explain our society's seeming inability to halt or reverse processes of environmental degradation. Human–nature dualism must be at the unchallenged center of our worldview if it is to explain the supposedly fierce resistance to the idea that humans are a part of nature. Contemporary practical resistance can only be explained by reference to the history of Western thought if that history speaks with one clear voice. To the extent that competing voices from within Western thought are heard today, then to that extent a theoretical bias in favor of a human–nature dualism could not explain the obstacles in our path to a more ecologically sound future.

Of course, no one would claim that *all* Western thought, or even all modern Western thought, is dualistic. Indeed, one could compose a familiar, albeit not agreed upon—list of "heroes" in the history of Western thought who are celebrated by contemporary environmentalist writers for their integration of human and natural concerns. Such lists have included St. Francis of Assisi, Spinoza, Emerson, Thoreau, Kropotkin, and Heidegger.[9] In order for the dualist interpretation to be consistent on this point, however, these Western thinkers must be regarded as lonely cries in the wilderness. They can be understood only as constituting a "minority tradition" with little impact (up to the present) on the "dominant paradigm" of the West.[10] Situating major figures in Western thought within the narrowly circumscribed boundaries of such a minority tradition ought, if they are convincing, to call into question the very existence of a dominant, dualistic paradigm in the first place.[11]

Earlier, I argued that a recognition of nature's role in human society is regarded by theorists of dualism as both novel and, on some level, uncontroversial. But how is it that an idea can be both a novel challenge to all existing ideas *and* uncontroversial? As I have illustrated here, an emphasis on a dualist interpretation of Western thought helps to resolve the apparent contradiction between these two claims. In the end, however, it cannot succeed.

To the degree that the ecological insight about humanity's interconnectedness with the rest of nature is uncontroversial, this is because ideas about human–nature interaction *do* resonate within contemporary Western societies. Yet such resonances call into question the overwhelming dominance of a single voice within Western thought, and hence the novelty of environmentalists' position. If Western thought were truly so dualistic, after all, then presumably most of us who are influenced by it should find the view that humans are constituted by nature to be ultimately *incomprehensible,* rather than more or less incontrovertible. That this position is comprehensible and even compelling suggests that it is not as novel or as threatening to existing thought as many imagine. This leads us to conclude that current thought is not as wedded to a human–nature dualism as presumed or argued. On the other hand, to the degree that elements of the position advanced by environmentalist thinkers are novel (including many arguments about the moral and political standing of natural beings and entities) and hence challenging to current thought, they are controversial.

Must the dualist interpretation be understood in the sort of "all-or-nothing" manner described here? Could it not be the case that those critical of the persistence of dualism in Western thought want to make the more limited point that nonhuman nature has been devalued, while some quality presumed to be unique to humans—reason, language, religion, soul, culture, subjectivity—has been elevated and valued? In this case, the dualist interpreters would not focus their argument on a failure to recognize the interconnectedness of nature, but instead on a failure to find intrinsic value in the nonhuman natural world. If this were the case, of course, then the evidence in support of the interpretation would be quite strong.[12] Many Western thinkers clearly do devalue the nonhuman world. Moreover, the preoccupation with finding some sort of intrinsic value in nature clearly is key to much of the writing by environmental philosophers. The dualist interpreter loses much, however, if she or he retreats to this more qualified position. Most significantly, she or he loses the presumption that to truly recognize natural connections is to embrace and value nature intrinsically. As a result of this loss, the dualist interpreter would be unable to assert that the transcendence of dualism is in any way noncontroversial, or clearly supported by the insights of ecological science. The ecological emphasis on interdependence and interrelationship so central to the interpretations considered above would no longer provide support for the antidualist position if dualism were understood in a way that is independent of an argument about this ecological conception of nature.

The dualist interpretation must, then, assert that the distinguishing characteristic of duality is whether nature is seriously considered in relation to humanity and human society at all, with the implicit assumption that if nature is considered in this manner, it will also be valued appropriately. Another way of stating this is that for the dualist interpretation to be coherent (though not, I have argued, compelling), the duality must go "all the way down." In itself, it cannot coherently advance the environmentalist positions it is intended to advance if the perceived division is one of greater or lesser normative valuation (of humans vs. nonhuman nature) resting upon a more basic understanding of human–nature interrelationship.

Deriving Politics from Nature

As I detailed in chapter 2, prescriptions for the future relationship between nature and politics often take on a derivative character among environmentalist thinkers. Thus, unlike dualism, approaches that seek to derive human politics and social order from a natural standard are not in themselves necessarily viewed as problematic. Instead, the problem is typically located in the particular conception, or conceptions, of nature that have played this role in the West.

What are the conceptions of nature that have been especially influential in Western thought? Despite the significant variety of conceptions that have been advanced over time,[13] there are two that have been especially important. The first, which is most closely associated with the natural philosophy of Aristotle, emphasizes teleology as central. Nature, from this perspective, is seen as offering a pattern of cosmic order in which natural objects and beings pursue their own proper ends. The second, which emerged with the scientific revolution in seventeenth-century Europe, offered a marked contrast to Aristotelian teleology and natural order. Perhaps most commonly associated with names such as Galileo, Descartes, and Newton, this conception of nature is typically referred to as mechanistic.[14]

Aristotle's teleological conception of nature is often seen as synonymous with his well-known anthropocentric claim that "plants exist for the sake of animals and . . . other animals exist for the sake of human beings."[15] He is explicit, moreover, in maintaining that political regimes "exist by nature" and that "man is by nature a political animal."[16] By combining the anthropocentrism of the first quotation with the political naturalism of the latter two, it becomes relatively easy for scholars to conclude that Aristotle's political argument is derived from an anthropocentric conception of nature. This conclusion invites the environmentalist rejoinder that Aristotelian politics is misguided because of its foundation in a mistaken conception of nature itself.

As I have noted earlier, the mechanistic conception that emerged with the scientific revolution is an especially familiar target of criticism among environmentalist writers. Central to these critiques is the argument that not only is such a mechanistic conception of nature at the core of modern science, but it is also at the core of modern social and political thought. For this view to be coherent, it cannot rely upon Descartes' strategy of divorcing humanity and the human mind from mechanistic nature. Instead, a mechanistic conception of nature must be viewed as one that can be and has been applied to the realm of humanity and human society. Here, Thomas Hobbes provides a seemingly more convincing model than his contemporary, Descartes.[17]

In her influential study, *The Death of Nature: Women, Ecology and the Scientific Revolution,* Carolyn Merchant argues that the mechanistic conception encouraged the objectification, and hence the conquest, of both women and nonhuman nature in a manner that could never have been legitimated by earlier "organic" conceptions that she identifies with Renaissance naturalism, and ultimately traces back to Aristotle.[18] A mechanical order of nature is one purged of life and inherent forces or principles. In its place, the mechanistic conception offers a view of inert nature composed of interchangeable parts and subject to externally imposed control and power.[19] Hobbes, in particular, is singled out in Merchant's analysis and critique. Not only did he hold a thoroughly materialistic, mechanistic philosophy of nature, but he:

further mechanized the cosmos . . . by reducing the human soul, will, brain, and appetites to matter in mechanical motion, and by transforming the organic model of society into a mechanistic structure.[20]

In Merchant's view, then, Hobbes unfolded the implications of the new conception of nature as they relate to individuals, society, and political order. Merchant offers abundant evidence that these implications were laden with gendered assumptions and were devastating in their effects on both women and nonhuman life. Her critique, however, does not rest upon this unfolding per se. Indeed, her analysis presupposes that moral, social, and political views will follow from one's conception of nature. Thus, for Merchant, this latter conception is best understood as the directive principle for a worldview that includes all the former views within it. The focus of Merchant's critique is on the new natural philosophy itself.

In *The Ecological Self,* Freya Mathews offers a similar assessment of the mechanistic conception of nature and of Hobbes's project. Mathews argues explicitly that the "social and ethical implications" of the mechanical, atomistic view that she labels "Newtonianism" are most accurately developed by Hobbes.[21] While Hobbes was far from the only one to do this, it is his ideas that represent:

the unsugarcoated implications that our culture has in fact absorbed and built into its normative structure. If Newtonianism did indeed reflect the world as it really is, then a Newtonian [Hobbesian] social system would indeed be natural, and to that extent legitimate.[22]

Mathews argues that a Hobbesian social system is not natural or legitimate, of course, but precisely (and only) because "Newtonianism does not reflect the world as it really is."[23] She interprets developments within twentieth-century physics as offering an alternative conception of nature that can now provide a foundation for a social system identified with the ideas of deep ecology, which she embraces.[24]

Merchant and Mathews are clearly critical of the conception of nature that emerged with the scientific revolution. What may be a bit less evident, however, is a key reason they are so critical. It is because they are convinced that a conception of nature can and does serve as a directive principle for human moral, social, and political organization and action. The effect of this conviction is that the politics of Hobbes's *Leviathan* appears to be derived from the mechanistic conception of nature that he embraces. As a result, these environmentalist critics focus on the conception of nature that he advanced, rather than the conclusions he claimed to have drawn from this conception.

The Allure of the Derivative Interpretation

A striking feature of the derivative interpretation considered here is that it views past Western thought in a way diametrically opposed to that of the dualist. How, then, can we make sense of the fact that these two radically different accounts have been developed by environmentalist thinkers seeking to advance the same, or at least very similar, ends? A discussion of some important structural and functional similarities in these otherwise conflicting views of the past can help to answer this question. Despite their great differences, each of these accounts can serve a similar role in justifying (and explaining the need for) an emphasis on an ecological conception of nature as the basis for social transformation in the future.

To interpret the history of Western social and political thought as deriving from the conception of nature dominant in each period is to view this history as consistent with the aspiration to a transformation of worldview described in chapter 2. While the introduction of an ecological conception of nature would lead (in the future) to a transformation of social and political thinking, this transformation would parallel those that have taken place over the course of Western thought. The seemingly most evident and explicit example of this is located in the seventeenth century, with the rise of a mechanistic conception of nature, and of modern philosophy and political thought.[25]

As novel and radical as the ecological conception is said to be, the derivative view of Western thought suggests that it is also profoundly consistent with the past to expect that this new conception of nature will transform our thinking about human relations. Moreover, because our current philosophy is regarded as a consequence of a particular conception of nature, and because this conception is now regarded as inaccurate as a result of the insights provided by ecology, it becomes logical to argue that our social and political ideas are outdated and in need of replacement. The derivative interpretation allows one to appeal to changing understandings of nature, largely as developed in contemporary science, as justification for a reorientation of social thought and practices. Although radical, the derivative interpretation sees this reorientation as consistent with the historical basis for such changes.

Moreover, and most noteworthy, there is no need to defend controversial social or political changes on their own terms here. Instead, only the new ecological conception of nature requires direct support. This appears relatively uncontroversial, if not absolutely conclusive and incontrovertible. Like the dualist interpretation, then, the derivative interpretation can also seduce environmentalists into believing that our appreciation for the importance of ecological interdependence can serve as sufficient justification for, and perhaps as an agent of, transformative social and political change.

For the derivative interpretation of Western thought to serve these purposes, however, it must meet certain important conditions and presumptions. Most important, this interpretation must be understood as an encompassing one. I identified the same precondition for the dualist interpretation in an earlier section of this chapter. For very similar reasons, the derivative interpretation offers little utility if it is advanced as merely an interpretation of some thinkers within the West, rather than the primary or dominant mode of developing social and political thought. After all, it is worth reiterating that environmentalist thought is ultimately concerned with guiding our *practices*. As such, a derivative account of the past can bolster the case for an ecological society in the future only if the former is understood as the way in which social practices have in fact emerged. If the history of Western thought reveals a pattern in which dominant social and political orders were shaped by influential political theories which themselves emerged from dominant conceptions of nature, then there would seem to be no theoretical obstacle for environmentalists to overcome. Instead, the challenge would be to advance an appreciation for the truth of the new ecological conception of nature.

A second important condition for the derivative account is that only one conception of nature be dominant in any particular era. It is central to the role played by this interpretation that it be possible to characterize the social systems of our age as fundamentally reflective of, say, a mechanistic conception of nature. The importance of this, once again, has to do with the relationship of theory to practice. It thus transcends the question of interpreting Western thinkers themselves. It is not enough to argue, for example, that Hobbes derived his Leviathan from a singular, mechanistic conception of nature. In order for this interpretation of Hobbes to address the practical interests of environmentalists, it must also be possible to say, with Freya Mathews, that Hobbes's view represents "the unsugarcoated implications that our culture has in fact absorbed and built into its normative structure."[26] In this view, our culture itself becomes a manifestation of a particular conception of nature. If conceptions of nature were regarded as less dominant and unified, then it would become necessary to examine and explain why one particular view was selected or highlighted. This sort of explanation, we have observed, would lead one away from a singular focus on conceptions of nature and instead to a consideration of the social and political perspectives that influenced such a selection process.

By identifying both the appeal and some of the presumptions behind the derivative interpretation, I hope to have raised some doubts about both the functional value and the coherence—and hence the validity—of this interpretation. Another way to initiate such a critique would be to advance an argument from the supposed "naturalistic fallacy" in which moral principles are said to be improperly derived from natural ones, or more generally "ought" derived from "is." This is not my intent here, at least in part because my goal is not to purify moral and political argument of natural ideas. Those whom I describe as dualist interpreters often criticize this goal, and while there are good reasons to believe that their interpretation is inadequate, I believe that they, like the interpreters that I consider in this section, are ultimately correct in their insight that we must seek to relate conceptions of nature and politics. And yet, the view of nature as a directive is unable to make adequate sense of how conceptions of politics, society, and humanity relate to nature. The question is thus how to identify or establish a relationship that does not regard nature as the standard without resorting to a dualistic rejection of any such connection. Considering nature and politics together in a way that is neither dualist nor derivative is a considerable yet vitally important challenge that will play a central role in subsequent chapters.

Conclusion

My first intent in this chapter was to distill the essence of the two dominant ways that environmentalist thinkers understand the relationship between nature, on the one hand, and humanity, culture, and politics on the other, in Western thought. Having identified these—the dualist and the derivative interpretations—I then sought to describe the foundations of these interpretations. By doing so, it should now be clear, my intent has been to build a critique of both. Although this critique begins to emerge in this chapter, it cannot really become clear until we look much more closely at the writings of some of the figures in Western thought who play key roles in the relationship that environmentalists so often discuss. This is my objective in chapters 4 and 5.

At this point, a bit of clarification may be valuable. While I criticize the oversimplifications and overgeneralization of both the dualist and the derivative interpretations, my intent is not simply to muddy the waters of environmental inquiry. Promoting the recognition that things are more complex than environmentalist thinkers often acknowledge is in itself no real accomplishment. Theoretical arguments should simplify complex relationships to a degree, in order to clarify the essence of this relationship and advance a particular argument. Yet by considering both of these interpretive poles, the limitations of each become most evident. We can conclude that neither the dualist nor the derivative interpretation is wholly accurate only by also acknowledging that neither is wholly *in*accurate. It is thus the partial insights of each that can point us toward a more insightful alternative.

Dualist interpreters are correct to argue for the recognition that humans and human society are inescapably interdependent with, and hence located within, nonhuman nature. It is because this seems self-evident to those writing on the subject that they often present it as offering a viable basis for widespread acceptance of ecological concern. However, only a view of humanity as part of nature would possess this quality of seeming incontrovertibility.

The role played by this constitutive conception of nature is, then, quite distinct from a natural standard. The latter might seek to promote values described as ecocentric, for example, on the grounds that a recognition of our interdependence and participation in nature demands that we attribute equal or at least significant value to (elements of) this nature. However, this natural standard is not a necessary consequence of the constitutive conception of nature.

In its most general sense, a constitutive conception of nature reminds us that humans are natural beings, subject to the same physical laws as the rest of nature. While our ability to create artifacts, history, and culture (of a certain magnitude) certainly distinguishes us from other natural beings, it does not exempt us from our physical interdependence with the rest of nature, or from biological needs and desires. It seems unquestionably true that, as Ted Benton has expressed it, "humans, as living organisms, depend for their organic well-being on their (socially mediated) relation to their ecological conditions of life."[27] The way we live our lives and the way we develop our culture cannot be separated from the natural world within which we are embedded.[28] Just as communitarian social and political theorists rightly emphasize that we truly are not self-constituting, atomistic, "unencumbered selves," but that we actually and necessarily develop within and respond to a particular social context,[29] so do environmentalist theorists rightly emphasize that we also develop within and respond to a natural context. Neither sort of theory should demand that we view context as determinative, but both require that we recognize it as constitutive.

Does the constitutive conception of nature lead us to an "ecocentric" view, or some other view in which nonhuman nature is regarded as having inherent or intrinsic value? Not necessarily. After all, crime and violence, war, poverty, and disease are important elements of the social "environment" for some individuals. However, we certainly would not want to argue that this obliges us to attribute *value* to such conditions. Both nature and society are necessary for the existence and the flourishing of human lives. Nonetheless, we certainly do not and should not value all elements of either the social or the natural environment that constitute who we are.

Given our dependence on the natural world for our existence, it seems more promising to suggest that a constitutive conception of nature could lead us to regard the protection of nature as having instrumental value. However, attributing instrumental value to nonhuman nature has the potential to be a tenuous basis in itself for protecting that nature. In some cases, there may be little about the constitutive conception that would limit a science fictionlike effort to *reconstitute* ourselves by transforming natural conditions, while still operating within the constraints of inviolable physical laws. The likelihood of such a transformation is not at issue here.[30] The point that I wish to emphasize is that no particular normative posture toward nature or toward particular types of natural beings or entities is a consequence of accepting a constitutive understanding of the role of nonhuman nature within human life and society. If we wish to advance a moral or political defense of nonhuman nature, we will need to do so independently of, and in addition to, an argument on behalf of the constitutive conception.

The relevance of the argument that I am advancing here may also be understood by considering its mirror image. We should recognize that by highlighting and perhaps even attaching privileges to certain qualities that are said to distinguish humans from the rest of nature, one need not be led to ignore or reject that nature's role in constituting who we are. It is true that many important Western philosophers have denigrated or failed to value nonhuman nature. However, this *normative* posture in relation to the nonhuman world is not equivalent to the assertion that these philosophers fail to account in a *constitutive* sense for nature or a natural context in their theorizing about human society and politics. In this latter sense, nature has frequently played a role in Western political philosophy. As a result, from an environmentalist perspective, the role of past Western thought is more ambiguous. The insight that can be obtained from a wide variety of these thinkers—even those with whom we ultimately disagree—becomes much greater. At the same time, the description of Western thought as a unified worldview to be transformed becomes far less coherent.

By disentangling the constitutive conception of nature from the critique of pervasive dualism in Western thought with which it is often linked, it becomes possible to recognize the importance of the former without being obliged to concur with the veracity of the latter. If we do so, then we can explore and evaluate the differing ways that nature and politics could be related, rather than construing the mere recognition of such a relationship as our primary task. We can begin to see that particular relationships may be more or less in line with our interests and values as environmentalists, as concerned citizens, or as community participants. This acknowledgment belies the belief that any simple recognition of nature could lead to the realization of a hoped-for ecological society. Conversely, it should also weaken any concern that an environmentalist agenda is inescapably antihumanist.[31]

At the same time that I wish to move away from viewing Western thought as essentially dualist, I also wish to urge us away from the trap of embracing the contrasting interpretation in which nature is seen as having provided a standard from which Western thinkers have derived social and political arguments. This characterization of the relationship fails to imagine that a particular conception of nature can be congruent with multiple political conceptions, and thus once again depends almost entirely upon the promotion of a new, ecological conception of nature as the basis for political and social change.

We must break the bonds imposed upon our vision of possible futures by rejecting the binary opposition of dualism and derivation as interpretations of the past. Doing so requires that we develop a more subtle understanding of this past; an understanding that can be fostered by careful attention to the arguments advanced by key figures in Western thought.

II
Rethinking Nature in Political Theory

4

Mechanical Nature and Modern Politics: The System of Thomas Hobbes

In this chapter, I examine the relationship between nature and politics that lies at the heart of Thomas Hobbes's philosophy. Doing so allows us to utilize Hobbes's thought to obtain critical insight into a relationship central to both political philosophy and contemporary environmental politics. I show that their roles are developed through a dialectical relationship in which Hobbesian politics shapes his discussions of nature in significant ways. This challenges Hobbes's own claim that his political philosophy is the distinctive product of a modern, mechanistic conception of nature—a claim that is central to many characterizations of Hobbes as the first distinctly "modern" political theorist. It also counters the contrasting interpretation in which Hobbes's "modernity" lies in his *rejection* of nature as a standard for political order. An explication of the tension between these interpretations is central to this chapter. It is important to emphasize here that each of these interpretations regards Hobbes as having destroyed "the dominant world-view through which his contemporaries surveyed man and political society,"[1] replacing it with the supposedly distinct and cohesive worldview labeled "modernity."[2]

This chapter has four main sections. In the first, I explore the meanings of "nature" and of "politics" in Hobbes's work. Next, I sketch the derivation of the latter from the former that he claims to have established in his theory, showing the importance of this claim to Hobbes and to our understanding and appreciation of his work. Despite Hobbes's claim to have derived politics from nature, a number of influential commentators have argued that in fact there is a disjuncture between his political and natural ideas. This dualist interpretation of Hobbes is the subject of my third section. In the final section, I argue that Hobbes was unsuccessful in maintaining the derivative relationship that he professes, and that the dualist interpretation suggests that such a failure was probably inevitable. This failure, however, should not lessen our interest in understanding the philosophical moves that Hobbes made in order to try to establish the relationship he envisioned. Moreover, we need not assume that dualism is the only alternative to a derivative relationship between nature and politics. I conclude that what Hobbes offers us is an implicitly dialectical relationship between nature and politics.

The centerpiece of Hobbes's unsuccessful effort to derive politics from nature can be found in his characterization of the *state of nature*. I show that for Hobbes, the state of nature serves as a mediating concept between his general conception of nature and his politics. Hobbes's use of the language of "nature" for this concept, however, clouds our appreciation of the role that it actually plays in his political philosophy. Most especially, it obscures the crucial influence of social and political conceptions on Hobbes's state of nature. I seek to demonstrate both the existence and the importance of this influence. It is this point that I develop in some detail and that is the key to my argument. Hobbes presents the state of nature as "natural," yet he incorporates a number of interpretations and judgments about politics into this "natural condition," while maintaining links with his broader conception of nature.

Definitions

"Nature"
Hobbes's conception of nature emerged from the intellectual ferment of the scientific revolution in which he participated. It was Galileo, he contends, who "was the first that opened to us the gate of natural philosophy universal."[3] Fundamental to the new natural philosophy, which Hobbes sought to elaborate, is "knowledge of the nature of motion."[4] A preoccupation with motion as key to unlocking the mysteries of nature was not an innovation of either Hobbes or Galileo, however. Indeed, the Aristotelian cosmology that the seventeenth-century natural philosophers rejected was also based on a core set of convictions about "the nature of motion." Aristotelian motion, or change, in nature was an indication of something striving to fulfill its natural potential or to find its natural place. Teleology

was thus central to the received wisdom about nature and motion prior to Hobbes's time. By contrast, Hobbes contended, motion must be understood strictly in terms of efficient, mechanical causation.[5] His conception focused on the inertial properties of motion and rejected any notion of final causes.[6] On these points, Hobbes was in agreement with the other preeminent natural philosophers of his era, from Galileo to Descartes, Gassendi, and Mersenne.

Whereas Aristotelian teleology emphasizes rest as the natural condition for the elements of the cosmos when they are in their natural place, the centrality of motion for the natural philosophy adopted by Hobbes cannot be overestimated.[7] In contrast to the mind–body dualism of Descartes, Hobbes presented everything as matter in motion, including our thoughts, senses, and will. His opposition to the Cartesian dualism is quite significant because of its seeming concurrence with the view advanced by many contemporary critics of dualism, as discussed in chapter 3. That Hobbes holds this view, however, poses a serious challenge to any effort to generalize about the prevalence of dualism among the major figures in Western thought. By advancing a monistic view, Hobbes seemingly avoids being subject to this critique while officially committing himself to a position of philosophical determinism.[8]

Nature understood in this manner becomes the singular, and inescapable, organizing principle of everything in the universe. As he unequivocally expresses it in *Leviathan,*

every part of the universe is body, and that which is not body is no part of the universe. And because the universe is all, that which is no part of it is nothing (and consequently, nowhere).[9]

If everything is understood as matter (or "body") in motion, then only the physical contact or pressure of matter can alter any other matter, a doctrine aptly characterized by one Hobbes commentator as "No change without push."[10] The only act that could be understood to escape this determinism would be the "first cause of all causes"—which we call God—that initiated the process.[11]

The antiteleological character of Hobbes's conception, like that of modern natural science in general, also suggests that we must dismiss any anthropomorphic understandings of nature. As historian of science Alexandre Koyré explains it, the philosophers of the scientific revolution appear to have substituted:

our world of quality . . . the world in which we live, and love, and die, [for] another world—the world of quantity, or reified geometry, a world in which, though there is place for everything, there is no place for man.[12]

This change prevents Hobbes from deriving normative standards of human behavior from his conception of nature in any simple manner.[13] However, Hobbes's determinism and his monism do suggest a strategy to overcome this difficulty; they appear to allow him to move effectively from talking about matter and motion to talking about human behavior.

The specific concept that serves this bridging function for Hobbes is "endeavour" ("*conatus*" in his Latin texts).[14] Defined by Hobbes as "*motion made in less space and time than can be given*" or "*motion made through the length of a point, and in an instant or point in time,*"[15] the concept allows him to characterize human cognitive and sensory processes as "motions of the mind,"[16] despite the lack of any visible movement. If human psychology is established as "natural" in this manner, and if this natural human psychology can be convincingly linked with a particular political order, then the definitive derivation of political theory from a proper conception of nature would be achieved by Hobbes. He would have overcome the antianthropomorphic proscriptions of the new science.

Hobbes equates his mechanistic description of human psychology with an unchanging human "nature." This is made possible by the firm link that he posits between human nature and nature in general, in the manner just outlined. Once humans are understood as they are "by nature," then (and only then) can reason recommend to us a series of "natural laws" that delineate how sovereignty and society should be organized, given this nature.[17] It is the derivation of these "dictates of reason," or "laws of nature," that Hobbes terms "the true and only moral philosophy."[18]

The conception of nature outlined here is nothing if not parsimonious. Yet its reach extends to everything in the universe. There is another important quality of Hobbes's nature, however, that must also be discussed. It is contentless. Nature defined as matter in motion tells us nothing about the character of that matter, or the direction or velocity of its motion. *Quality* is eliminated from the new conception of nature espoused by Hobbes and the other natural philosophers of his time.[19] As Hobbes expresses it, whatsoever accidents or *qualities* our senses make us think there be in the world, they are not there, but are seemings and apparitions only. The things that really are in the world without us, are those motions by which these seemings are caused.[20] [emphasis added]

While Hobbes's argument allows him to move from a consideration of the physical world to human nature, as mentioned earlier, the exclusion of content or quality from his understanding of "nature" dramatically limits his ability to speak about either the physical or psychical content that we also generally subsume under the appellation "nature."

One important consequence of this conception of nature is that it actually allows a great deal of space for Hobbes's nominalism. Nominalism—the conviction that abstract concepts have no referent in the external world but exist merely as the names attributed to things by humans—might easily appear to be in conflict with the naturalistic strand of Hobbes's argument that I have been developing to this point. After all, in his nominalistic vein, Hobbes denies the existence of universals; "this word *universal* is never the name of any thing existent in nature," a claim that seems in conflict with his position on the universality of body as the central reality of nature.[21] Yet it is precisely the absence of content in his conception of nature that provides the basis for the artifice of naming—and artificial human creation more generally—in the first place. Thus the two need not, at least always, be in conflict.

Hobbes defines as "accidents" all those qualities of matter that I have suggested would fill in the content of his conception of nature. He is insistent upon distinguishing the two. Accident is *"the manner by which any body is conceived"* or *"that faculty of any body by which it works in us a conception of itself."*[22] Hence, he concludes, "the object is one thing, the image or fancy is another."[23] Because these "accidents" are conceptualized within our minds, and are not qualities found in nature itself, they are subject to the distinctly human practice of naming. Thus, as one scholar has explained it, for Hobbes, "conceptions in the mind bridge the signs of language and objects in the external world."[24] This understanding of the relationship between nature and nominalism for Hobbes does not fully address the basis for his definition of nature itself.[25] It does, however, show how central his particular conception of nature is to the delineation of an expansive sphere that he terms artifice, established through language, while still allowing for the consistent, although contentless, naturalism throughout his theory.

The argument for the compatibility of a particular conception of nature and the artifice of nominalism in Hobbes will only be convincing if it can also address the relatively more familiar and seemingly very real contrast that Hobbes presents between the chaotic, conflictual state of nature and the ordered, peaceful Leviathan that epitomizes his understanding of artifice. Here, the artificial seemingly offers humans an escape from "nature," where the latter is famously presented as offering us a life that is "solitary, poor, nasty, brutish, and short."[26] Where nature is understood as Hobbes's description of the "state of nature," or "natural condition of mankind," it is this "nature" that poses the problem for which artifice is presented as the necessary solution.[27]

What is the connection between Hobbes's articulation of the "natural condition" as a framework for understanding human relationships to each other and to the world around them and his embrace of the mechanistic conception of nature? I argue in a subsequent section that the former cannot properly be understood as a consequence of the latter within his theory, although it is important to his systematic project to suggest that it can be.

"Politics"

While references to nature and the natural abound in Hobbes's writings, "politics," the other concept central to my inquiry, is not a term of distinction for Hobbes. To consider this in a meaningful way then, we must first explore those related terms that Hobbes does use distinctively and see what they can tell us about how he conceives politics. Two terms that tell us much about Hobbesian politics are "sovereignty" and "commonwealth."

The social contract, as described by Hobbes, is the singular and definitive means of escaping from the "natural condition" that he describes as a "war . . . of every man against every man."[28] Through this mutual agreement, "commonwealth" is created. It is, says Hobbes,

as if every man should say to every man *I authorise and give up my right of governing myself to this man, or to this assembly of men, on this condition, that thou give up thy right to him, and authorize all his actions in like manner.* This done, the multitude so united in one person is called a COMMONWEALTH.

The "man, or . . . assembly of men" is the one whose acts everyone has made themselves the author of,

to the end he may use the strength and means of them all, as he shall think expedi-
ent, for their peace and common defence. And he that carrieth this person is called
SOVEREIGN.[29]

Sovereignty is thus presented as the answer to the problem posed by the
state of nature. Sovereignty encompasses the whole of human artifice and
is manifest as an all-or-nothing alternative to our natural condition. In
contrast to thinkers for whom civil society or private life can or does
exist apart from the state, for Hobbes they can only be understood as
encompassed within the state. Commonwealth or civil association itself
results only from the creation of sovereign power. The exercise of sovereign power becomes synonymous with politics, properly understood.
There is no other possible location for political activity within his scheme.

The equation of politics with sovereignty means that the former is seen
as extremely broad in scope. The same equation, however, also excludes
much of the content that we might otherwise consider within a description of politics. While the ends of politics are intended to be directed—
as described in the definition of the sovereign—to the pursuit of peace,[30]
decisions about the form and process of politics and political decision-
making are dismissed from theoretical consideration and designated as a
mere "difference of convenience."[31] Hobbes does express a clear preference for monarchical government,[32] but his equation of politics with
sovereignty prevents him from incorporating this preference into his philosophical system.

Similarly, Hobbes cannot rule out disagreement or debate within the
(collective) body of a (nonmonarchical) sovereign. Differences of opinion
are natural, according to Hobbes, both between individuals and even
within the same individual over time.[33] While he is insistent in seeking
to eliminate these differences as a challenge to the sovereign power, he
offers us no reason to believe that they can be eliminated within the sovereign body itself. Nonetheless, by focusing on politics strictly as a unified
sovereignty, Hobbes can have little or nothing to say about processes by
which disagreement might be more or less likely to be channeled into
effective decision-making.

Having presented Hobbes's understanding of the two concepts central
to my inquiry here, in the following sections, I consider the relationship
between them. It is there that both the most interesting and the most
problematic aspects of Hobbes's theory emerge.

The Derivative Account

Contemporary thinkers have often ascribed a great deal of blame for our current environmental dilemmas to the mechanistic conception of nature developed during the seventeenth century. As we saw in chapter 3, Carolyn Merchant and Freya Mathews offer especially pointed criticisms of Hobbes as the source of these dilemmas. While Hobbes was far from unique in his advocacy of nature as mechanistic, Merchant and Mathews argue that he is distinguished by his effort to derive a new understanding of political power and social order from this view of nature.[34] He is also distinguished, they argue, by his forthright, "unsugarcoated" appraisal of the social and political consequences that can be derived from this nature.

In his latest book, William Ophuls summarizes this position as unequivocally as it can be:

[the new science] swept away the old worldview and the social order based upon it . . . [and] led directly and immediately to the new political thought. . . . Hobbes, whose explicit aim was to reconstruct politics in the light of the new Scientific understanding, . . . made building the commonwealth into a matter of engineering—that is, of applying mechanistic principles to social life. He therefore began *Leviathan* with an explanation of the basic principles of *mechanistic* philosophy precisely in order to establish the basis for an *individualistic* theory of politics.[35]

Since Ophuls, like Merchant and Mathews, rejects the mechanistic conception of nature, the consequence appears to be a rejection of Hobbesian politics as well.

There is good reason to question whether Hobbes was as successful in molding his psychological and political theory to conform to his natural philosophy as these authors presume. There is no reason, however, to believe that he would have been unhappy or uncomfortable with their presumption. If this relationship is not truly derivative, however, then at least some of the criticisms advanced against Hobbes's conception of nature may turn out to be misdirected. Understanding why this is so requires outlining Hobbes's philosophy as he presented it and considering why the links between the different sections of his system appear so important to him.

Leviathan is the best-known and most-read work by Hobbes. It reflects his mature ideas about politics and human psychology. It is not, however, always the most revealing statement of Hobbes's overall philosophical ambitions. Hobbes presented his Latin trilogy—*De Corpore, De Homine,* and *De Cive* ("Body" or "Matter"; "Man"; and "Citizen")—as the complete statement of his philosophical system long before the completion and publication of all three sections. This reveals most clearly his view of the relationship between nature and politics.[36] In *De Corpore,* a work published after all of his major political treatises, Hobbes describes the intended relationship between his conception of nature and his politics or "civil philosophy." Discussing how demonstrable knowledge can be produced in various fields, Hobbes posits that:

he that teaches or demonstrates any thing, [should] proceed in the same method by which he found it out; namely, that in the first place those things be demonstrated, which immediately succeed to universal definitions. . . . Next, those things which may be demonstrated by simple motion. . . . And after these, the motion or mutation of the invisible parts of things, and the doctrine of sense and imaginations, and of the internal passions, especially those of men, in which are comprehended the grounds of civil duties, or civil philosophy; which takes up the last place.

Lest there be any question of his argument here, Hobbes goes on to emphasize that "this method ought to be kept in all sorts of philosophy";

such things as I have said are to be taught last [civil philosophy], cannot be demonstrated, till such as are propounded to be first treated of [natural philosophy], be fully understood.[37]

Hobbes outlines a hierarchical and derivative sequence in which all subsequent philosophy is ultimately dependent upon our fundamental understanding of nature itself. In this formulation, conclusive knowledge of "civil philosophy" is contingent upon prior knowledge of "sense and imaginations," which itself is dependent upon knowledge of "simple motion," knowledge that can be traced back to a set of first principles about body or matter. That this is Hobbes's conviction about the proper relationship between knowledge of nature and of politics is also reinforced by the organization and ordering of the three sections of his Latin philosophical treatise itself.

And yet a reader of Hobbes's political philosophy might remain skeptical. After all, there is only a relatively brief discussion of Hobbes's mechanistic conception of nature in *Leviathan,*[38] while *De Cive* offers no such discussion. Yet, as noted, these works were published years before his major work of natural philosophy, *De Corpore.* Moreover, in the preface

to *De Cive,* Hobbes justifies the appearance of this work prior to the two
sections that would logically precede it by explaining that because his
civil philosophy is "grounded on its own principles sufficiently known
by experience, it would not stand in need of the former sections."[39] All
this would appear to be at odds with his claims, described earlier, about
the inherent unity of his philosophy and its dependence upon an understanding of nature.

Hobbes, of course, was aware of this apparent conflict within his philosophical work. From his point of view, the conflict was not real, however,
because at least two paths could lead to civil philosophy. Hobbes described the first as the "synthetical method," in which a commonwealth
is constituted on the first principles of matter and motion. The second is
the "analytical," and begins with everyday questions and resolves them
back to necessary first principles.[40] Hobbes suggests that his predominant
method in his political treatises is the "analytic," in the sense that first
principles are said to emerge from introspection and experience. *In principle,* however, Hobbes maintains that civil and moral philosophies *could*
be derived from his conception of nature itself.[41] This conviction is crucial. It is only this conviction that allows Hobbes to claim that his politics
have an unshakable foundation in nature. Introspection, after all, could
differ in quite significant ways among individuals. It is only if Hobbes
can establish the naturally determined subject of our introspective reflections that he can be confident of their universality. Without this natural
foundation, Hobbes offers no other basis for assuming that humans are
alike (enough) for him to build his political theory upon introspection
alone.

Hobbes viewed himself as the originator of (true) civil philosophy,
which he brazenly claimed to be no older than his own work, *De Cive.*[42]
Yet such a claim is nonsensical if we regard introspection alone as the
basis for his theory. Certainly many had been introspective in their political reflections before Hobbes. What allows him to claim to have origi-
nated the true civil philosophy is his development and utilization of what
he took to be both the method and substance of the new natural philosophy or science.[43] The connection between method and substance, in this
regard, is quite significant.

Hobbes, following Galileo, regarded geometry as the language of nature itself. It is only, after all, if one regards the underlying "nature"

of nature as fundamentally geometric or mathematical that a geometric method is appropriate for the pursuit of natural philosophy or science. As Galileo argues,

Philosophy is written in this grand book, the universe, which stands continually open to our gaze. But the book cannot be understood unless one first learns to comprehend the language and read the letters in which it is composed. It is written in the language of mathematics, and its characters are triangles, circles, and other geometric figures without which it is humanly impossible to understand a single word of it . . . "[44]

Similarly, in *Leviathan*, Hobbes argues that

senseless and insignificant language . . . cannot be avoided by those that will teach philosophy without having first attained great knowledge in geometry. For nature worketh by motion, the ways and degrees whereof cannot be known without the knowledge of the proportions and properties of lines and figures.[45]

The application of the same method to the study of humans and the human creation of political order reflected, for Hobbes, the conviction that the underlying "nature" of these realms is, despite all other differences, congruent with that of the world that humans exist within. In the absence of such a conviction, one might just as easily suggest that another, radically different, sort of reasoning or calculation is appropriate to the realms of human behavior and human creation, a suggestion that cannot be found in Hobbes's political treatises.

Hobbes's system can now be seen. Unified sovereignty is a political necessity that is legitimated by his argument that it is the true political philosophy because it is the only one derived from the true conception of nature now revealed by natural science. Of course, the key legitimating function played by the concept of nature should not suggest that Hobbes celebrated or legitimated what he took to be the natural condition within his political theory.

For Hobbes, the *condition* of nature is inconsistent with human livelihood and what he terms "commodious living."[46] As one scholar notes, "[t]he overall view . . . was not so much a natural order as it was a natural disorder."[47] This, of course, is famously expressed in his description of this condition as a war of all against all. It is equally relevant to Hobbes's understanding of the relationship of humans to their physical environment. "The end of knowledge," says Hobbes, echoing Francis Bacon, "is power."[48] The utility of natural philosophy is made evident, according to Hobbes, by its ability to lead to other varieties of human artifice, including engineering and technology, that allow for human power and control over "disorderly" nature. In Hobbes's words:

the greatest commodities of mankind are the arts; namely, of measuring matter and motion; of moving ponderous bodies; of architecture; of navigation; of making instruments for all uses; of calculating the celestial motions, the aspects of the stars, and the parts of time; of geography, &c. By which sciences, how great benefits men receive is more easily understood than expressed.[49]

This application of natural science to the physical environment is dependent upon the creation of the most crucial form of artifice in Hobbes's system: sovereignty. In the absence of Hobbes's sovereign, the Leviathan, he argues that we would also lack all those forms of artifice that allow humans to significantly modify or control the natural world. This is expressed in his portrait of life in the state of nature as one with

no culture of the earth, no navigation, nor use of the commodities that may be imported by sea, no commodious building, no instruments of moving and removing such things as require much force, no knowledge of the face of the earth, no account of time, no arts, no letters, no society. [50]

The two points are closely connected. For Hobbes, it is only by first overcoming the condition of nature through the creation of a sovereign via the artifice of the social contract that we can expect to make any significant progress in the pursuit of "commodious living." The creation of sovereign power in turn allows for the modification of the physical environment through the advancement of science and technology, as well as the delineation of property rights. Again, artifice is the key to this change.

And yet, to reiterate a central point, by premising his theory upon "nature," Hobbes sought to establish a firm and unwavering foundation for it. *This* "nature" must be understood as unchanging for him. The conception of nature as motion does not allow for alteration by human artifice. If it did, then Hobbes's philosophical determinism would be undermined, and all his theories would appear to be up for grabs. More generally, if nature as matter in motion were itself subject to human change, then the universal explanatory power of the new sciences would be severely undermined. The opening sentence of *Leviathan* defines "nature" as "the art whereby God hath made and governs the world."[51] While human artifice is said to *imitate* nature, Hobbes could not characterize it as the *transformation* of nature, so defined.

The condition that nature puts us in, however, is changeable—and indeed must be changed—for Hobbes. Maintaining the distinction between "nature" and the "natural condition" thus becomes crucial for Hobbes. The distinction separates the latter, understood as a problem or challenge posed to us by nature, from the source of this challenge. If Hobbes was unable to maintain the distinction consistently, then either artifice would be impossible (and hence sovereignty, technology, etc., would have to be reconceptualized as somehow natural) or nature could no longer be understood as foundational, since it would be negated by the social contract and the creation of artifice. Either possibility would seriously undermine Hobbes's system. I return to this dilemma in the next section of this chapter.

Although it may be a less familiar element in his system, Hobbes also utilizes his conceptions of both nature and the state of nature as a means of delineating and circumscribing sovereignty itself. As one Hobbes scholar has observed "[t]he central theme . . . is the unity of the state; it is neither the citizen's liberty nor the total state."[52] In order to assert this theme, Hobbes must establish the basis for credible constraints upon both alternatives, "citizen's liberty" and the "total state." It is important to Hobbes to characterize these constraints as "natural." If he succeeds, then all that would seem to be needed is for him to accurately identify them for us. By contrast, if they are not natural in this sense, then he would be obligated to persuade us—or the sovereign—of the importance of establishing these particular constraints. Given Hobbes's unwillingness to conceive of any realms of artifice outside of sovereignty, moreover, nature seems to be the only possible location for constraints or boundaries to the scope of sovereign power.

Hobbes seeks to avoid the specification of conditions in which citizen revolt (or any other source of independent, humanly created power) might be legitimated, while still characterizing sovereignty in a way intended to preclude totality. In a well-known passage in *Leviathan*, Hobbes refers to the sovereign's laws as "hedges" intended "not to bind the people from all voluntary actions, but to direct and keep them in such a motion as not to hurt themselves."[53]

This metaphor of the civil law as hedges—borders—reflects a particular conception of the ends of sovereign power. If the sovereign's task is solely to establish hedges, then the possibility of either more grandiose or more pernicious ends is ruled out. Hobbes's sovereign advances the "common good," but only insofar as this good has been redefined and limited by natural law. According to Hobbes, the fundamental natural law is to "seek peace,"[54] and it is only this radically circumscribed end that is consistent with sovereign power.

By defining the *telos* of sovereignty as limited to peace through the maintenance of "hedges," Hobbes's limited sense of the proper role for politics rules out a number of other possible applications of state power. Visions, for example, of a therapeutic state, an educative state, or a state engaged in other forms of social engineering are rendered inconsistent with this characterization.[55] This inconsistency reflects the apparent impossibility of achieving such visions, given Hobbes's understanding of nature and human nature.

Nature itself, we must recall, is matter in motion, and this motion is understood to be distinctly nonteleological. This lack of natural teleology and the corresponding lack of ends prescribed by human nature are the basis for Hobbes's fundamental law of nature. Provided that Hobbesian subjects recognize this "law," further state efforts to impose any more specific ends upon them are by nature bound to fail.

"Constraint" may be an inappropriate word to characterize this sort of Hobbesian limitation on sovereignty, if we understand this word to suggest some sort of human-imposed or -enforced limitation, and thus some sort of human power that could exist above or outside of the supposed sovereign. These, I have noted, are ruled out. Yet the sovereign *is* limited by this formulation, and limited by (Hobbes's conception of) nature. Such an argument is not inherently inconsistent for Hobbes, since the artifice that is sovereignty is not opposed to the nature that is matter in motion, and the latter is as real within the commonwealth as it was within the state of nature.

A second, related, limitation on sovereignty within Hobbes's theory is a reflection of his conception of human nature. More properly, perhaps, it is a reflection of his emphasis on a fixed and determined human nature, which itself is a manifestation of nature in general. No contextual, cultural, or historical change can be understood to alter human nature. If it could, then the requirements for a political theory would also be subject to change. It is only because this nature remains unchanged despite radical changes in structural context that a universal solution to the warring condition of the state of nature is possible. It is precisely this sort of solution that Hobbes maintains he has offered.

For Hobbes, "men's belief and interior cogitations are not subject to [the sovereign's] commands."[56] This statement is presented not as a normative one, but as descriptive of the nature of our beliefs. This can only be the case if it is not possible for a Hobbesian sovereign to alter our nature, or "denature" us. If it were possible, then by altering our beliefs it might be possible to transform us into persons for whom such a sovereign would no longer be a necessity. For example, we might become persons for whom self-preservation is no longer regarded as the primary (natural) good. Of course, all people do not always regard self-preservation in this manner. Hobbes acknowledges at times that some are motivated by values (religious and otherwise) that they hold higher than their own preservation.[57] He argues against such motivations, however, on the grounds that they are unnatural, hence irrational.[58] They are unnatural because "life itself is but motion," and hence any commitment that conflicts with or threatens to end this motion is irrational because it fails to recognize the utter finality of this end.[59] Yet if a sovereign were able to alter our nature, then even this argument in favor of self-preservation could be undermined. More generally, if we are not understood to have a determined, unchanging nature, then the whole notion of the "self" to be "preserved" could be threatened by the sovereign power.

This claim for the natural inviolability of human selves not only allows for the possibility of a universally applicable political theory, it can also restrain the potential ambitions of the holder of sovereign power by excluding the possibility for success of actions intended to reshape or reconstitute our selves. It is worth emphasizing here that in attributing to Hobbes an argument about the inviolability of human nature by human artifice, I am not suggesting that he held what we might term a Kantian view of an autonomous self. While the Hobbesian and Kantian views may find commonality in their conception of a human subject resistant to social or political influence, they differ markedly in the basis for this resistance. Hobbes's subjects are not autonomous, they are determined. It is because they are said to be determined by nature, however, that he can maintain their internal independence from sovereign attempts to alter their (human) nature. Again, Hobbes relies upon his conception of nature to maintain the distinction between a unifying and a totalizing model of politics—precisely the sort of distinction that he would otherwise be unable to make.

The Dualist Interpretation of Hobbes's Thought

In the preceding section, I sought to demonstrate that Hobbes presented his political philosophy as derivative of and dependent upon his conception of nature and that it was at least reasonable for him to have done so. Moreover, I have tried to describe some of the key strategies and arguments utilized by Hobbes in presenting his philosophical system, and the centrality of his natural philosophy to this presentation. I have not, however, tried to argue that Hobbes's derivation is either wholly coherent or convincing. It is neither.

In fact, a number of influential twentieth-century Hobbes scholars have argued that there is a disjunction between his political and natural ideas. They have argued not merely that Hobbes does not successfully derive his politics from his nature, but have gone further to argue that there is actually *no* substantive relationship between the two. The Hobbes that emerges from this interpretation is one whose political ideas developed from a consideration of a denatured humanity and whose stance is distinctively antinatural. In this sense, their interpretation of Hobbes— which I call the dualist interpretation—appears as a mirror image of the derivative account that they reject.

Leo Strauss presents an especially significant account along these lines. He explicitly denies any meaningful connection between Hobbes's conception of nature and natural philosophy on the one hand, and his political philosophy on the other. Strauss argues that one must choose between what he terms the "naturalistic" (i.e., derivative) and the "humanistic" (i.e., dualist) interpretation of Hobbes's political philosophy; he chooses the latter.[60] For Strauss, Hobbes's political thought is founded entirely upon his humanistic convictions, convictions that Strauss regards as wholly at odds with Hobbes's natural philosophy. Presenting a stark interpretive dichotomy, he argues that

the student of Hobbes must make up his mind whether he is going to understand Hobbes's political science by itself or whether he is going to understand it in the light of Hobbes's natural science.[61]

He clearly sides with the first of these understandings. He argues forcefully that

the conception of nature which Hobbes's political philosophy *presupposes* is dualistic: the idea of civilization presupposes that man, by virtue of his intelligence, can place himself outside nature, can rebel against nature. . . . The antithesis of nature and human will is hidden by the monist (materialist-deterministic) metaphysic, which Hobbes *teaches* . . . [but] which is not only not needed for his political philosophy, but actually imperils the very root of that philosophy.[62]

For Strauss, as well as his followers among political philosophers, the disjunction between nature and politics that he characterizes as central to Hobbes's approach is the distinguishing mark of modern political philosophy.[63] Strauss's normative take on this issue is also compatible with the environmentalist critique of dualism, as considered in chapter 3. Both regard the separation of nature and politics as problematic. While there is no indication of environmental concern in Strauss's work, and while he clearly rejects any deterministic naturalism based upon the new science, his criticism of this separation is a reflection of his commitment to a standard of natural right as the basis for an appropriate political philosophy—a standard that continues to be espoused within environmentalist thought today.

Another influential scholar of Hobbes's political philosophy, Michael Oakeshott, advances a similarly dualistic interpretation while adhering to a normative perspective somewhat more common among recent philosophers and political theorists (and at odds with that of many environmentalists). Oakeshott characterizes Hobbes's *Leviathan* as the "supreme expression" of a tradition of thought centered around the concepts of "Will and Artifice," concepts that he contrasts with the previously dominant tradition of political philosophy centered around "Reason and Nature."[64] In this manner, Oakeshott also rejects any connection between Hobbes's politics and his natural philosophy. For Oakeshott, however, will and artifice are concepts to be celebrated. Unlike Strauss, he exhibits no real worry about the demise of nature as a standard.

Despite this normative contrast, however, the interpretive emphasis remains the same. Hobbes, and by extension the other modern political theorists that followed him, are characterized as having driven a wedge between the philosophical inquiry into nature and the philosophical inquiry into politics. Nonetheless, because Hobbes's political philosophy is presented as the fountainhead for the tradition of thought that regards politics and society as wholly artificial or non-natural, it appears to provide confirmation for those environmentalist authors seeking to trace the roots of a way of thinking that allows for the exploitation or destruction of the natural world.[65]

One important basis for the dualist interpretation of Hobbes's thought has been located in the genealogy of his ideas. If it can be convincingly shown that Hobbes's political ideas were developed in some depth before his attention turned to natural philosophy, then it would seem logical to conclude that his political ideas were not derived from his natural ones, and in fact are independent of them. It is this strategy that is especially central to Strauss's early work on Hobbes, in which he examines the "basis and . . . genesis" of Hobbes's political thought.[66]

On the one hand, the ability to advance this argument is significantly eased by Hobbes's precise identification of the point in time when he first encountered the text of Euclid's geometry. According to Hobbes, it was this encounter, in 1629, that provoked his interest in and study first of geometry itself, and subsequently of natural philosophy.[67] Prior to this time, Hobbes's studies were primarily humanistic and historical. Since Hobbes was already forty years old at this point, this seems a good reason to suspect that, in the words of a biographer, Hobbes's political philosophy "had its main lines fixed when he was still a mere observer of men and manners, and not yet a mechanical philosopher."[68] On the other hand, there is no clearly convincing evidence of his political views prior to this age; all of his well-known political treatises were published after he was in his fifties.

Strauss's project requires him to trace key elements of Hobbes's political philosophy back to earlier writings, especially the introduction to his translation of Thucydides (which was published in the same year that he discovered Euclid). If Strauss can do this successfully, then a convincing case might be made for the claim that Hobbes's political philosophy grew initially out of his humanistic studies, rather than out of his natural philosophy.[69]

Strauss's thesis received some encouragement recently from the seemingly authoritative identification of three anonymously published discourses as the work of the young Hobbes.[70] Strauss had suspected Hobbes of writing these discourses (among others),[71] but it is through innovative computer analysis that Hobbes's authorship now seems to have been es-tablished.[72] What can we learn about Hobbes's ideas from these early essays, assuming that his authorship is established? Certainly, we can learn that Hobbes's political thought did not emerge, *sui generis,* from his conception of nature. It would seem clear that the former was not in fact strictly derived from the latter by Hobbes. There are many differences between the young Hobbes of the so-called *Three Discourses* and the mature Hobbes of *Leviathan.* Nonetheless, there is enough commonality to also allow us to conclude that something other than a derivation of politics from nature has occurred. In this limited sense, then, the Straus-sian thesis seems significantly bolstered.

There are, however, significant limits to the ability of a historical analy-sis to establish the relationship between Hobbes's conceptions of nature and of politics. Implicit in Strauss's claim is the premise that if he can convince us that Hobbes did not derive his politics from his nature, then he will also have established the truth of Hobbes's "humanistic," or dual-istic, interpretation. The existence of common elements in Hobbes's early (pre-Euclid) political ideas and his more mature ideas on the subject would be enough to undermine certain claims about a derivative relation-ship. The existence of differences in his ideas over this time must then be attributed either to a simple maturation of his (independent) political thinking or to an effort on Hobbes's part to camouflage his true, early, ideas in order to make them more palatable to the audiences for his pub-lished works.[73]

Notice once again the opposition that Strauss's interpretation of Hobbes creates. Either Hobbes derived his politics from his natural ideas or he drew them from strictly humanistic sources. Once the former possi-bility is discounted, only the latter dualistic interpretation remains. The acceptance of this interpretation is then a necessary precursor to the de-velopment of explanations for the existence of differences in his ideas over time, as sketched in the previous paragraph.

If the initial opposition is correct—if Hobbes's philosophy must be either derivative or dualistic—then Strauss's interpretation seems valid. However, what if the opposition is not correct, or if we admit other pos-sibilities for the relationship between nature and politics in Hobbes's philosophy? In this case, an argument that calls into question a deriv-ative relationship would offer only limited insight into Hobbes's actual conception of the nature–politics relationship. In this case, we could consider the possibility that Hobbes's conception of nature is highly relevant to his political philosophy, even if the latter was not derived from the former. Ultimately, there is no a priori reason to believe that the dichotomy central to Strauss's interpretation is correct. As a result, a dualistic account of Hobbes's philosophy must appeal to more than just its genealogy to establish its validity.

There is much in Hobbes's political philosophy itself that encourages a sort of dualistic interpretation, even apart from its origins. It is this that Oakeshott and other interpreters have in mind when they contrast Hobbes's emphasis on will and artifice with an earlier centrality of reason and nature. Whereas the derivative interpretation of politics in Hobbes's theory begins with his mechanistic conception of nature, the dualist interpretation is on its firmest ground when it begins with a consideration of his state of nature.

If we accept Hobbes's account of the state of nature, or "natural condition," as adequately describing his understanding of nature and the natural, it becomes relatively easy to arrive at an interpretation that is radically at odds with the derivative account sketched in the preceding section. Indeed, if one were to limit one's consideration of Hobbes to a reading of his explicitly political texts, then this interpretation may be the most straightforward and exoteric of all. Certainly many undergraduates who read selected portions of Leviathan in an introductory course in political theory come away with a thumbnail sketch in which Hobbes deems "nature" conflict ridden and bad, while political order is characterized as an artificial, human construction and inevitably good. If pressed, such an undergraduate might further assert that Hobbes rejects the idea of nature and natural law as a standard against which we can measure a polity, and that he advances an argument that concepts contain no objective or universal meanings, but are instead defined through words (nominalism).[74]

Hobbes does indeed advance positions like those attributed to him by our hypothetical undergraduate. The overall implication of these positions is that human creation is somehow distinctly non-natural and that there is an unbridgeable divide between it and nature. Given this reading of Hobbes, environmentalists seeking to trace and evaluate the origins of dualistic and antinaturalistic thinking in the West are wholly justified in identifying Hobbes as a key villain. The question that should emerge by this point, however, is whether such a dualistic reading can be sustained, especially given an awareness of Hobbes's contrasting, derivative arguments that we considered earlier. Even if we isolate his political writing from his natural philosophy, as Strauss explicitly recommends and Oakeshott at least implicitly endorses as well, we are left with the question of whether Hobbes offers us the moral and political conceptions necessary to allow the antinaturalist interpretation of his politics to cohere.

Oakeshott has already suggested to us what is required for Hobbes's political philosophy to fit within the dualist and antinaturalist framework. It is adequate conceptions of will and artifice. Moreover, these conceptions must be untainted by any connection with his natural philosophy. On this score, Hobbes cannot be said to have succeeded. Artifice, we have seen, contrasts sharply with Hobbes's conception of the state of nature. For it to contrast equally with his understanding of nature itself, Hobbes would have to offer an understanding of the human activities necessary to create this artifice as also non-natural. However, he is unable to provide a description of human agency that is consistently distinct from his naturalism. "*Will,*" Hobbes asserts in chapter six of *Leviathan,* is merely "*the last appetite in deliberating.*" This definition "makes no action voluntary; because the action depends not of it, but of the last inclination or appetite."[75] To anyone influenced by the derivative account of Hobbes's politics, this definition is wholly unsurprising. For the dualist interpretation, however, it becomes inescapably problematic. Hobbes asserts later in *Leviathan* that wills "make the essence of all covenants."[76] Yet if "will," for Hobbes, is not non-natural, then the covenant that results in the "artifice" of the commonwealth cannot be said to be distinguished from this conception of nature either.[77]

In the end, then, the dualist interpretation of Hobbes forces us to acknowledge, more than the derivative account did, the absolutely key and highly favorable role that he attributes to concepts of will, artifice, and nominalism within his political philosophy. And yet, the strategy utilized here to do this requires us to remove Hobbes's ideas about nature and natural philosophy from consideration. This strategy is not successful because Hobbes does not offer us a way of cleanly separating his political thought, dependent as it is upon concepts of will and artifice, from the understanding of nature that, for him, shapes these concepts in such important ways.

Hobbes's politics cannot be fully understood in isolation from his natural philosophy and so it should be clear that an appreciation of his effort to relate nature and politics is central to an understanding of his philosophy. This understanding can be furthered only by examining the links that the scholars who advance a dualist interpretation find unimportant. However, when we examined these links, it also became clear that Hobbes's politics fail to fully cohere when the relationship to his natural philosophy is characterized as a derivative one. In the final section of this chapter, I suggest an alternative formulation in which the nature–politics relationship is viewed in a more open-ended and dialectical manner.

Hobbes's Dialectic between Nature and Politics

By first acknowledging Hobbes's ambition of deriving politics from nature, we may begin to see why it was so important to his project. Certainly, he sought a firm foundation for his political theory. Equally important, Hobbes's conception of nature was itself a deeply held conviction that was both the impetus for and the product of a significant portion of his life's work.[78] And yet, Hobbes can only define his Leviathan as "an artificial man"[79] because he suggests, and at first seems to maintain, a clear boundary between nature and artifice. Nature is warlike and chaotic, while artifice is peaceful and orderly.

It would seem that Hobbes was trying unsuccessfully to have it both ways. His conception of nature as matter in motion is monistic and inescapable. This conception prevents artifice from being situated on the opposite side of a conceptual boundary from nature—an obstacle that dualist interpreters overcome only by ignoring or rejecting his natural philosophy. If artifice is to be possible within Hobbes's theory, it must be contrasted with a "nature" other than the one articulated within his natural philosophy. To make sense of this, we must sort out the ambiguities in the diverse appeals to nature made by Hobbes and his interpreters.

Human artifice requires the introduction of a new concept—the natural condition, or state of nature—into Hobbes's theory. It is at this point, however, that the derivative character of the system breaks down.

Hobbes's argument for the transcendence of the state of nature cannot be reconciled with its allegedly natural character. While his state of nature is influenced by his conception of nature, the former cannot be solely defined by the latter. It must also be shaped by his unacknowledged judgments and interpretations about the character and scope of politics (and artifice) itself.

The state of nature is a concept that bridges the gap between nature and politics only by partaking of some qualities from both of these concepts. The consequence of accepting this interpretation is striking. We can no longer regard the "natural condition of mankind" as a direct or logical manifestation of nature itself, for Hobbes. As a result, Hobbes's politics can no longer be understood as the response to a problematic condition that must be conceived as truly natural. The political arguments that Hobbes develops cannot be understood as derived from a prior conception of nature and human nature. Instead, they have been formed through a dialectic between these conceptions and his political presuppositions.

In *Leviathan,* Hobbes describes the "natural condition of mankind" as that where there is "no power able to over-awe them all."[80] The key to this definition is not any conception of nature or human nature, but a particular form of artifice understood as "power able to over-awe them all." This is Hobbes's understanding of sovereignty. Similarly, in another passage in this same discussion, Hobbes speaks of how nature "dissociates" humans.[81] It would seem to be easy to criticize Hobbes here simply on the grounds that the language of dissociation seems to presume a historically prior form of association. This would be deceptive, however, since Hobbes is fairly consistent in maintaining that his description of the state of nature is a thought experiment, rather than an intended description of an actual historical period.[82] What is most important and revealing about this passage is not that Hobbes fails to describe the state of nature as temporally prior to human association, it is that he also fails to present it as *logically* prior. To conceive of the state of nature as filled with individuals who have been dissociated, then, raises the question: "Dissociated from *what?*" The answer, for Hobbes, is "from that all-encompassing artifice, *Leviathan.*" It is important, then, that the state of nature emerges as the negation of this prior conception of sovereignty, rather than as a foundation for it, or even for any more general notion of artifice.[83] Striking though this conclusion is in light of Hobbes's philosophical ambitions, it appears inescapable if he is to avoid the untenable position of defining artifice in contrast to his monistic conception of nature itself.

In order for the artifice/state of nature distinction to avoid collapsing into an artifice/nature dualism, Hobbes must establish the distinction in the first pair strictly upon differences in structural conditions, rather than appealing to any distinctive "naturalness" of the state of nature. Especially in his mature formulation in *Leviathan,* this is what he seems to do. Given the existence of human passions, it is the process of human reasoning *within the particular structural context* of the state of nature that is described as a primary cause of the war of all.[84] In this description, it is not the existence of human passions, reasoning, or human nature per se that accounts for the state of nature. After all, these are understood to be universal. They are just as real within Hobbes's England, and within his ideal vision of "commonwealth," as they are in his state of nature. Human passions in themselves, therefore, cannot be held responsible for the state of nature, since the transcendence of this condition involves no alteration in our passions themselves (only in their objects).

Similarly, the artifices of the social contract and of Hobbesian sovereignty are based on human reason. Reason is, however, "no less of the nature of man than passion."[85] Reason, which can lead us to "natural law," is thus the foundation for artifice. The latter can therefore be no less (or more) natural for Hobbes than the state of nature. Both can best be understood as structures within which unchanging nature and human nature act in particular, although quite distinct, ways.

In the previous three paragraphs I sought to detail a coherent relationship between artifice and the state of nature to which Hobbes generally adheres. Before moving on, it is worth emphasizing the uniqueness of this relationship and the definitions that it relies upon and, more important, those that it rules out. As we have seen, a nature/artifice dualism must be ruled out by Hobbes. Yet if the artificial is *not* defined in relation to some prior conception of nature, then the term itself appears to lose its familiar meaning. Hobbes does offer an alternative "nature"—the state of nature—with which we can contrast artifice. Here it turns out that it is the artifice of sovereignty itself that allows Hobbes to define the state of nature in the first place, again undermining any natural or prior basis for identifying what is or is not "artificial" in his theory.

Hobbes goes further than merely suggesting a distinction between a condition in which the structures created by humans dominate (artifice), and one defined by the absence of these structures (state of nature). Also crucial to his political theory is the move to equate human artifice in general with absolute sovereignty in particular. This move is at times oblique despite (or perhaps because of) its importance. Of course, if Hobbes were able to make us forget about the importance of this move by conflating artifice and sovereignty, then he would only need to convince us of the need for artifice rather than the particular, unified, and hierarchical form of artifice—sovereignty or Leviathan—that he is advocating.

Indirectly, Hobbes does conflate the two. As discussed earlier, he defines the state of nature as the absence of absolute sovereignty. Escaping the state of nature, then, would appear to require the creation of its opposite. Yet the particular sovereign character of this creation is already evident in the definition of the problem. In the end, artifice is a category defined by Hobbes as synonymous with his notion of sovereignty. This definition reflects his convictions about the requirements for social and political stability. It presents these convictions as conclusions, however, when in fact they are the premises of his argument itself.

Historically, Hobbes directed his political arguments for unified sovereignty against those that sought to provide justification for autonomous realms of authority, whether that realm was the church, parliament (before its ascension to sovereign power), common law, universities, or experimental science.[86] Since we have seen that Hobbes presents none of these as viable in the state of nature, they all must be recognized as human artifacts. As a result, it should not be possible for Hobbes to conflate his argument for sovereignty with a position that views artifice itself as necessary to overcome the war of all against all that characterizes the state of nature. And yet, in more than one way, he does. Perhaps the most surreptitious of these ways has already been analyzed. By defining the state of nature as the absence of sovereignty, as Hobbes does, human artifacts in general appear indistinct from sovereignty in particular.

An examination of another important step that reinforces the equating of artifice and sovereignty allows me to further explore the influence of

Hobbes's conception of politics on his portrait of the state of nature. This examination will also reveal the distance between the actual relationship of the central concepts in Hobbes's political theory and the systematic relationship that he posited, but could not maintain.

Hobbes makes a significant attempt to ground the argument for unified sovereignty in his nominalism. As a nominalist, he repeatedly emphasizes the absolute importance and temporal priority of defining terms. This emphasis is consistent with his reliance upon geometry, which itself utilizes a nominalist method in which theorems are deduced from a set of first principles defined by humans. Sovereign authority thus appears necessary in order to provide a singular, unified set of definitions. Hobbes's justification, however, relies upon his argument that discrepancies in the definition and valuation of terms (especially good and evil) are *themselves* a primary cause for conflict and war in the state of nature. As he expresses it in *Leviathan*,

divers men differ . . . [in] what is conformable or disagreeable to reason in the actions of common life. Nay, the same man in divers times differs from himself, and one time praiseth (that is, calleth good) what another time he dispraiseth (and calleth evil); *from whence arise disputes, controversies, and at last war.*[87] [emphasis added]

According to this argument, it is terminological or conceptual disagreement itself that is the cause of physical conflict. As expressed in *De Cive*, humans are "so long in the state of war, as . . . they mete good and evil by diverse measures."[88] Certainly, Hobbes also talks about conflicting passions and desires as a cause of war in the state of nature.[89] However, the argument quoted above is not specifically linked with one about conflicting passions. It is presented as an independent claim that the "fact" of nominalism, which leads to varied notions of the good, is itself inherently conflictual in the absence of a sovereign to overcome it. Without a separate argument that the diverse objects of our passions are necessarily in conflict, however, this argument cannot be convincing. If it were, then Hobbes's observation that individuals differ within themselves in their valuations over time would suggest that each of us is regularly at war with our own self. As one scholar puts it,

One can imagine that, starting from this conception of the sovereign's law as "common reason," and looking backwards to the antecedent time, he was driven to see it as the reign of conflicting individual reasons. But, considered in the general framework of the system, this conflict cannot but be interpreted as the theoretical expression of a clash of more fundamental desires.[90]

Interpreting conceptual disagreement as a mere "theoretical expression," which relies upon claims about conflicting desires within the structural context of the state of nature, may offer a more logical explanation of Hobbes's argument (although such claims still assume the existence of conflict). As a result, we might be tempted to write off the former as simply an enthusiasm, suggesting that perhaps Hobbes drew out the potential implications of his account of nominalism or conceptual disagreement farther than was appropriate.

Hobbes's argument about conceptual disagreement as a cause of war does more for his theory, however, than this suggestion to set it aside recognizes. It is only in the context of this argument that he can maintain his sovereign as definitional of terms and meanings. By presenting sovereignty in this manner, Hobbes excludes the possibility of allowing politics to include any sort of debate or struggle over meanings themselves. An argument merely for the "fact" of nominalism, and thus differing meanings (which Hobbes develops far more extensively in his work), does not of itself lead one to equate sovereignty with all artifice.

As I have already noted, Hobbes recognizes that differing nominal interpretations are commonplace even within a single individual over time. Yet this need not lead us to overcome this commonplace condition. Why? Because such differences are not necessarily conflictual. We are able to overcome some differences within ourselves and with others without the imposition of power from above. That this is not likely to be the case all the time, or in all cases, suggests the importance of forms of power able to resolve or contain disputes.

In itself, however, Hobbes's conviction about the existence of nominalism cannot do more for his argument than this. It is linking this "fact" with the very real problem of conflict that Hobbes's political theory intends to overcome that leads to this equation of nominalism with the need for sovereignty. After all, if debate over meanings is a cause of war, then it certainly could not continue to exist within Hobbes's commonwealth. Without support for this particular causal connection, however, Hobbes appears to have once again read back into his state of nature conditions that are only properly understood as a reflection of his convictions about artifice and sovereignty.

A second problem for the attempt to use nominalism to argue on behalf of sovereignty is the fact that nominal definitions themselves are quite clearly a human creation. Of course Hobbes recognizes this. The implications, however, are significant. For the artifice of language to be presented as a cause of war in the state of nature, it must be logically prior to the creation of sovereignty. This necessary introduction of artifice into the state of nature raises doubts about the very possibility of a form of sovereignty that can encompass all artifice because it further undermines the distinction between this latter concept and the state of nature itself.

Hobbes's political theory relies quite heavily upon a characterization of the state of nature that is unable to provide him with the sort of foundational support that he sought from it. Yet, as I discussed earlier, Hobbes also develops important elements of the contours and boundaries of sovereignty in relation to his conception of nature itself.

The nonteleological, mechanical, and deterministic qualities of Hobbes's conception of nature are clearly important to his theory. We have seen that they appear to allow him to dismiss or ignore the possibility of using artifice to change nature or human nature itself. Yet it is the lack of content or quality in Hobbes's conception of nature that encourages such a large role for artifice, and that should dramatically weaken our confidence that the existence of a natural determinism can actually preclude a radical alteration in the content of "nature." What are our ends? What does our world look like? This content can be transformed or altered through the structures of artifice, even though nature in the abstract—matter in motion—cannot. However, it is in the content that many of our concerns about nature lie, whether with threatening forms of social control or with agents of ecological destruction.

Thus, Hobbes's conception of nature appears to recede even further within his theory. While at first we saw it withdraw somewhat as a result of the distinction between nature and the state of nature, here we can see it retreat further as a result of the distinction between abstract and concrete understandings of nature. This distinction may be better appreciated by considering it in relation to the arguments about natural constraints on sovereignty discussed earlier in the chapter.

Hobbes's conception of nature tells us that humans have no natural ends, because the natural world of matter in motion that determines us has no *telos*. Yet, from this nature, he wishes to portray politics, or sovereignty, as having no end other than peace. At best, however, the relationship between these two claims is an analogy.[91] Hobbes's nature cannot provide a constraint that would prevent the sovereign (understood as an artificial structure) from ordering or organizing human ends even though they are not so "naturally." The support that Hobbes can claim for his effort to restrict the ends of sovereignty is in his depiction of the desire for "commodious living" as a natural pursuit secured only in this manner, and his understanding of natural law as the basis for ensuring this pursuit.[92] And Hobbes's natural law, once again, relies heavily upon his portrait of life in the state of nature—a portrait that we now see cannot convincingly be characterized as distinctively natural.[93]

It is largely as a result of what Hobbes defines as the "desire of such things as are necessary to commodious living, and a hope by their industry to obtain them"[94] that human manipulation of the nonhuman natural world takes place. At the same time, this definition of "commodious living" emerges as a key restriction on the power of the sovereign. Once we see that the restricted end of sovereign power is not an inevitable consequence of Hobbes's conception of nature, however, then we can also see that the particular ways in which the sovereign encourages, discourages, or directs human interaction with the material world are not inevitable or unalterable. Commodious living, for Hobbes, is a construction that defines our relation to the material world, yet seemingly does so outside the realm of sovereign power. However, the absence of teleology in Hobbes's nature does not, or should not, prevent us from considering and questioning how commodious living is pursued, or even, more particularly, from doing so via sovereign political institutions.

We have seen that Hobbes's conception of nature, and especially his determinism, also seem to prevent the possibility of "denaturing" human selves. Again, however, a lack of content is attributed to this determined (human) nature. As a result, an argument that Hobbesian subjects cannot be denatured by their sovereign becomes a purely formal claim that remains compatible with a wide variety of efforts on the part of the latter to restructure or reorient human activities, beliefs, or ends. While our nature, so defined, remains unchanged, our lives are potentially subject to an array of tremendously powerful influences. And it is these latter influences that are of greatest consequence in shaping our own, inevitably qualitative and content-dependent, conceptions of who we are and what we seek.

If there are constraints on sovereignty that remain in Hobbes's system, then, they must be understood as ones that might result from the arguments or rhetoric that he employs. They may or may not be persuasive to us. Surely, however, they could not be construed as natural by Hobbes, nor should they be by us. Indeed, Hobbes seems to fear the arguments and rhetoric of elites as an especially pernicious threat to the sovereign he advocates.[95] Yet, in the end, he cannot succeed in turning to nature as a means of constraining sovereignty. More generally, rooting his political theory in his natural philosophy does not allow Hobbes to escape the very real world of politics and political disagreement.

Conclusion

This chapter's investigation began with, and was instigated by, the argument (advanced both by Hobbes and many interpreters) that Hobbes's political thought emerged from the new conception of nature produced by seventeenth-century science. This mechanistic conception radically transformed human understanding of the cosmos and of our place and role in it. Just as cosmology was being affected by these developments, so was epistemology. The method by which knowledge and understanding of the world could or could not be obtained was equally the subject of change during the period that Hobbes wrote. All of this has led to the suggestion, and often the explicit argument, that a new worldview emerged from the new conception of nature. Such a worldview would have clear and decisive consequences for our understandings of politics, society, and ethics, among other subjects, consequences that Hobbes seems to develop.

My study of Hobbes's arguments in this chapter led to the conclusion that if we wish to adopt (temporarily) the language of worldviews, then Hobbes's cannot be identified with his conception of nature itself. Instead, Hobbes's worldview—the one that has clear consequences for social and political thought—is best located in his conception of the *state of nature*. This latter conception is adequately understood only as the product of a dialectic between his preexisting conceptions of both nature and politics. Used in this way, the term "worldview" itself becomes less serviceable. It becomes far less of a fixed and long-standing framework for human thought and action than as otherwise conceived.

By examining the relationship between these categories central to Hobbes's thought, I do not simply seek to reveal the many ambiguities and contingencies that emerge. Hobbes, after all, has the reputation of being among the most systematic thinkers in Western political philosophy—a reputation deserved in many respects. His failure to consistently maintain a derivative relationship between categories of nature and politics should therefore begin to suggest to us the great difficulty, and perhaps impossibility, of doing so. A recognition of this failure undermines the claims of naturalness and universality that Hobbes and many others use to bolster their political arguments. It forces us to evaluate these arguments within particular contexts and with respect to the particular ends we might seek.

The dialectical relationship between Hobbes's conception of nature and his politics that I articulated here may or may not have been envisioned by him. Certainly, the systematic character of his philosophy would lead one to suspect he would have been well aware that key elements of his argument were inconsistent with the overall project that he proposed. Yet, I believe an examination of Hobbes's intentions in this regard would likely be fruitless and almost certainly inconclusive. Moreover, for the purposes of this study, the question of intent is largely beside the point.

What is not beside the point is the relationship between nature and politics revealed in Hobbes's writings. Certainly it is important that the determined, inertial, nonteleological character of Hobbes's conception of nature influences his description of the state of nature and hence his political argument. Yet it is equally true, and important, that a conception of politics or sovereignty that encompasses all artifice also shapes this state of nature. By mediating between his conceptions of nature and politics, then, Hobbes's state of nature blurs the boundary between nature and artifice in a manner that destroys any lingering confidence we may have that these spheres could have been understood and delineated before his development of a political theory.

5

Natural Ends and Political Naturalism?
Understanding Aristotle

I believe that scarce anything can be more absurdly said in natural philosophy than that which is now called *Aristotle's Metaphysics;* nor more repugnant to government than much of that he hath said in his *Politics;* nor more ignorantly than a great part of his *Ethics.*
—Thomas Hobbes[1]

If we seek a clear contrast to his own conceptions of nature and politics, Hobbes strongly suggests, we can do no better than turn to Aristotle. While Aristotle's authority and influence in the universities was waning by the midseventeenth century when Hobbes wrote the above words, it still constituted the conventional wisdom against which Hobbes and others directed many attacks. In Aristotle's work, therefore, we should expect to find not only a clear alternative to the conceptions of nature and politics developed by Hobbes, but also an interpretation of these conceptions that has been uniquely influential in Western thought. To the extent that the mechanistic conception of nature is perceived as a source of our contemporary environmental dilemma, moreover, in Aristotle's thought we may find elements of a more attractive alternative. His nature is not the "dead" mechanism of Hobbes, but is instead a realm filled with bodies that have an inner source of motion and change. The whole made up of these natural bodies appears to be ultimately ordered and harmonious; a purposive nature that seemingly ought to be both respected and valued.

Nature, for Aristotle, is a "good householder" whose orderly ways ought not to be disrupted without good cause.[2] This perspective on nature can be found throughout much environmentalist thought, as familiar (although problematic[3]) appeals to the "balance of nature" remind us.

Indeed, the term "ecology" itself was coined with explicit reference (like the much older word "economy") to the Greek *oikos,* and the notion of householding that it connotes.[4] Most important, Aristotle's nature is teleological. Here, too, we find much that is in accord with contemporary environmentalist thinking. In his book *Respect for Nature,* for example, Paul Taylor argues that viewing organisms as "teleological centers of life" is of central importance.[5] Similarly, Val Plumwood has argued that "[a]n important part of obtaining alternatives to a mechanical paradigm . . . is the reinstatement of teleology as important and irreducible."[6] Not all manifestations of Aristotle's ideas of nature, however, are so amenable to contemporary environmentalist thought. His well-known comment in the *Politics* that "one must suppose both that plants exist for the sake of animals and that the other animals exist for the sake of human beings" is frequently panned as extraordinarily anthropocentric by contemporary writers, and in a number of cases is presumed (mistakenly, I argue) to be the central doctrine of his natural philosophy.[7]

If a conception of nature can establish a worldview from which social and political answers can be derived, Aristotle's thought is likely to offer an especially clear and important example or model of this relationship.[8] His politics, after all, is avowedly natural, in contrast to Hobbes's emphasis upon its artificiality. In fact, as I discuss later, some version of a derivative relationship between nature and politics is central to the way Aristotle has been frequently understood. Aristotle's conception of teleology in nature is viewed by many as the basis for his claims that man is a "political animal" and that the *polis,* or political community, itself is natural.

This interpretation of Aristotle's politics lends credence to the belief among environmental writers that a change in worldview (represented here by Aristotle's natural teleology) is key to the realization of an "ecological" society. Even those writers who focus on Aristotle's "anthropocentric" passage, noted earlier, as the key to his philosophy of nature often regard his politics as having been derived from this view. In this case, however, environmentalist writers typically use this understanding as a basis for rejecting Aristotelian politics. Regardless of his view of nature, it should come as little surprise that I argue here that a closer look at Aristotle's conceptions of nature and politics reveals that he does not and cannot derive the latter from the former.

Not all interpreters, however, regard Aristotle's politics as natural and his view of the nature–politics relationship as derivative. Indeed, a second important school of thinking about Aristotle asserts almost precisely the opposite. Here, Aristotle is regarded as crystallizing a view of politics as a realm of freedom and artificial construction, in which freedom and artifice are defined as the absolute antithesis of necessity and nature.[9] In this view, Aristotle's *polis* is both conceptually and physically separated from his *oikos*. By identifying the latter as the realm in which all activities connected with "bearing and caring for life"[10] are located, Aristotle attempts to justify his well-known restriction of women, children, slaves, and others who engage in these activities to the household. Crucially, this second school of interpretation also argues that he circumscribed from the realm of politics all the issues that emerge from householding. Concerns related to agriculture, economics, gender relations, natural resources, and other areas tinged by their connection to the household and to "nature," are said to be excluded from Aristotelian politics. Hannah Arendt, in a few pages of her book *The Human Condition,* offers the seminal interpretation of Aristotle along these lines.[11]

Influenced by Arendt's interpretation, others have concluded that Aristotle "seeks to establish the *polis* as a transcendent and self-sufficient association."[12] The characterization of the *polis* as transcending the life process and nature more generally becomes central to an interpretation in which Aristotle is understood as a theorist—and often a preeminent theorist—of a dualism between nature and politics. Once again, I argue that this understanding is inadequate and unconvincing.

This chapter adheres closely in form to the structure of chapter 4. Before exploring the relationship between Aristotle's conceptions of nature and politics, I first clarify his understanding of these two conceptions. The following section sketches Aristotle's conceptual understanding of nature and of politics. After that, I consider Aristotle as a theorist who derives his political thought from his conception of nature. I consider the arguments of some scholars who accept or advance such a derivative interpretation and reconstruct the key steps of such an argument in Aristotle's writings. Next, I discuss the dualist interpretation of Aristotle, seeking to make sense of this contrasting argument as well. Finally, I develop an alternative interpretation of Aristotle in which I argue that a crucial ambiguity in his discussions of the political community prevents a derivative account of his thought from being convincing and under-mines the dualist interpretation in which the political community has transcended nature. As a result (as in the case of Hobbes), I conclude that while Aristotle's politics clearly is related to and reliant upon his conception of nature, this relationship is much more adequately charac-terized as dialectical rather than derivative. It is to Aristotle's definitions of nature and politics that I now turn.

Definitions

"Nature"

The Greek *physis*, rendered consistently as "nature" in English transla-tions, is a term of distinction in Aristotle's philosophy. It is not used hap-hazardly, nor is it simply equated with the totality of all things. To appreciate Aristotle's understanding and use of this term, it will be helpful to turn first to his natural philosophy and his biological writings, in which "nature" plays an especially prominent part. In this way we can grasp the role and importance of Aristotle's emphasis on teleology in nature, the relationship between nature and human nature, and the implications of these views for an overall vision of order and harmony in the cosmos.

A central contrast for Aristotle, described in his *Physics,* is between nature and artifice. The distinguishing characteristic of things that "exist by nature" is that each has "within itself a principle of motion" and of rest.[13] "Motion" as Aristotle uses it here is more appropriately identified with the broader concept of change, since motion (for us) typically con-notes a change in location, while Aristotle also associates the concept with changes in size and quality.[14] Natural things have an inner source of change and stability and "a source of movement within the thing itself is its nature."[15] Products of art, by contrast, lack any such innate impulse or source of change.[16] While they are composed of natural materials, a house, a bed, or an article of clothing come into being and change only as a result of the actions of an external source—human creativity. The external source of change must itself be natural (have an internal source of change) or at least must ultimately be traceable to a natural source, one with its own internal ability to change. This condition is met for those objects that are a creation of humans, since humans themselves are natural in this Aristotelian sense.

Although Aristotle presents four causes of natural change (traditionally translated as material, formal, efficient, and final[17]), it is the last of these that he most closely associates with nature itself.[18] Nature, Aristotle maintains, "is the end or that for the sake of which."[19] That "for the sake of which" a body strives, moves, or changes is properly understood as its final cause, end, or *telos*. If change is not externally imposed upon the body, then it is natural and the change is the reflection of an underlying natural teleology inherent within it. This view of natural teleology or purposiveness permeates Aristotle's writing and underlies his analyses. He describes natural teleology as a key principle that "we are accustomed constantly to use for our scientific investigation of nature," and which "we must take for granted . . . in all nature's work."[20] Inherent within this perspective is a presumption in favor of natural ends, so defined. Death, after all, is at the end of all life, but it is clearly not *an* end or *the* end in the sense that Aristotle speaks of teleology. By definition, the end "for the sake of which" a natural body strives is what is best—not what is last—for that body.[21]

Understood in this way, Aristotle's natural teleology must be distinguished from a view that equates nature with necessity. The end of a natural body is what is best. Inherent in the notion of "best" is the suggestion that things could be otherwise. While Aristotle does not deny the existence or importance of causally necessary relationships, he does not equate these with nature.[22] A natural thing's internal principle of change is manifested, Aristotle asserts, by the fact that such change happens "invariably or for the most part."[23] Although this formulation does not reject the possibility that a natural change might happen "invariably," it locates the basis of a change's naturalness elsewhere. A tendency—a "for the most part"—is a reflection of a thing's nature, but such a tendency is not manifested everywhere or always. "[T]he tendency in each is toward the same end," Aristotle asserts, "if there is no impediment."[24] The nature of an acorn is to become an oak tree, Aristotle suggests, so if there are no impediments, oak trees (and only oak trees) will result from acorns. There are, however, all sorts of contingent circumstances that might prevent this outcome; while it is "best," the natural outcome is clearly not a necessary one.[25] Indeed, in his discussion of artifice in his *Nicomachean Ethics*, Aristotle makes a point of distinguishing it from necessity and from nature, making it clear that the latter two are in no way the same thing.[26]

Teleology is not only central to Aristotle's own conception of, and investigations into, nature, for later thinkers it is also perhaps the single most controversial and oft-rejected element of his science. Because the rejection of teleology was so central to the scientific revolution of the seventeenth century (including Hobbes) and because the thinkers of this period and after have often attributed silly views to Aristotle on this subject, it is important to be clear about what his conception of nature as teleological does and does not mean.

Quite commonly, Aristotle has been accused of anthropomorphizing nonhuman nature. Treating natural bodies as having ends, it is argued, is to treat them as having intentions, something that seems nonsensical in the absence of consciousness and thus misguided in the study of nonhuman nature. What Aristotle has done, this argument suggests, is to falsely project intentionality onto natural bodies, thus developing causal explanations that are vacuous. The claim that change is the result of a body's striving to fulfill (or "actualize") its potential is said to offer little insight once we recognize (which Aristotle is presumed to have failed to do) that this sort of striving cannot reflect any conscious intention.

Teleological arguments need not be based on anthropomorphism, however, and it is unconvincing to argue that Aristotle must have viewed nature in this sort of misguided manner.[27] If he did not, however, then what does teleology mean for Aristotle? One possible answer is that he merely meant to suggest that we read nature *as if* it had ends or purposes. This answer, however, overcomes the critique of anthropomorphism only by radically downplaying, if not eliminating, any significant role for natural teleology. Here, teleology is reduced to metaphor for Aristotle, with the language of final causes interpreted as a mere colloquialism that allowed him to write in a way that would be easily grasped by his contemporaries. Yet this is unconvincing or at least incomplete, given the clearly prominent and distinctive place that Aristotle gives to the concept in his writings.

Another possible interpretation of Aristotle's teleology suggests that even if he did not attribute intention or consciousness to individual (natural) bodies, he did attribute these qualities to "nature" as a whole. In this view, nature has been reified as a singular entity that can possess qualities apart from those of particular natural bodies. Certainly, there are points when his discussions of nature do appear to be consistent with such an interpretation. Ultimately, however, this transcendent, Godlike conception of nature is not Aristotle's. As I discuss later, Aristotle always identifies nature first with the characteristics of particular bodies and particular "natures"; any discussions of nature as a whole should be understood only as an aggregation of these particulars.

It seems most convincing to read Aristotle as arguing that ends actually do exist for natural bodies, but that these are not to be equated with conscious intentionality. What he does not seem to offer is a clear answer to the question of how these ends are advanced in the absence of intentionality, leaving the sort of unsatisfying explanation that has prompted the alternative interpretations considered to this point. Nonetheless, he is confident that the regularity found in processes of natural change can only be adequately understood in this way.[28]

While not inherently anthropomorphic, then, Aristotle's natural teleology can allow for a relatively smooth transition between human and nonhuman nature. This is critical to any argument about the relationship between nature and politics. All natural bodies are characterized by an inner principle of change and this change is naturally directed toward an end. In humans, the movement toward an end may in fact be characterized as conscious and intentional. In this manner, Aristotle can recognize the uniqueness of human nature. At the same time, human behavior remains natural in this view. Indeed, in at least one place in his writings, Aristotle refers to man as "natural in a higher degree than the other animals."[29] It is characteristic of this conception of nature that while humans may be distinguished by their ability to create artifice and to act voluntarily, this in no way lessens their naturalness. As one recent commentator has argued,

What Aristotle terms "voluntary" acts . . . are precisely those actions which arise "naturally," that is, as the result of a principle of movement that lays within the agent.[30]

What we saw earlier with reference to Hobbes's theory is therefore also true of Aristotle's. Humans and human nature are recognized by both thinkers as inescapably a part of (the theorist's conception of) nature itself. The distinctive human capacity for reason and speech (*logos*) is situated within, not contrasted with, Aristotle's understanding of *telos*.

While an appreciation for this continuity between nature and human nature is important for the argument developed later, it is at least as important to appreciate the overall picture that emerges from Aristotle's conception of nature. Nature is strikingly heterogeneous. As I have already noted, it is more proper to say that for Aristotle, singular "Nature" exists only as the aggregation of the particular natures of particular kinds of natural objects. Aristotle often makes use of the (singular, unified) term "Nature," in such characteristic expressions as "Nature does nothing in vain," or "Nature always acts with a purpose." Yet, as one influential commentator argues convincingly, these views are best understood "not as a transcendental principle but as a collective term for the natures of all 'natural bodies' working harmoniously together."[31] This conception of nature has important implications for the manner in which study and understanding of it can be achieved.

No universal method would be appropriate for the study of such a distinctly heterogeneous nature. Natural change is specific to particular kinds of natural bodies and so can only appropriately be studied in a qualitatively specific manner. This particularity leads Aristotle to emphasize that each type of natural body will have a corresponding natural *place*—a place or particular context in which terrestrial bodies will rest, or in which celestial bodies will remain in perfect motion.[32] This concept of place is distinct from a generic conception of *space,* where the latter is conceived as a universal, homogeneous realm such as that envisioned by geometers.

In explicit contrast to the Pythagoreans, members of Plato's Academy, and other mathematically oriented predecessors, and in equally clear contrast to the science later developed by Galileo and followed by Hobbes, Aristotle rejects the notion of a mathematical study of nature. In their quest for universal forms and methods, he argues, the mathematicians lose sight of the particular, qualitative manifestations of birth, growth, and other sorts of natural change.[33] Since Aristotle's investigation of nature is fundamentally a search for causes of change, these qualitative manifestations are of the utmost importance.[34]

When all particular natures are aggregated, a picture of a naturally ordered whole begins to emerge. It is for this reason that Aristotle is described as viewing nature as composing the *cosmos:* a harmonious, ordered and finite understanding of the universe, in which everything has a natural place. Alexandre Koyré expresses this overall picture of Aristotelian nature well:

Whole, cosmic order, and harmony: these concepts imply that in the Universe things are (or should be) distributed in a certain determined order; that their location is not a matter of indifference (neither for them, nor for the Universe); that on the contrary each thing has, according to its nature, a determined "place" in the Universe, which is in some sense its own. A place for everything, and everything in its place: the concept of "natural place" expresses this theoretical demand of Aristotelian physics.[35]

"Politics"

Aristotle's conception of "politics" can only be understood by a consideration of both his *Politics* and his ethical writings, in particular the *Nicomachean Ethics*. Since the relationship between this conception and Aristotle's conception of nature is the focus of the following sections of this chapter, his well-known pronouncements early in Book One of the *Politics*, that the *polis* "belongs among the things that exist by nature" and that "man is by nature a political animal," will only be noted here.[36] Of course, these claims are integral to Aristotle's theory, but the way in which they are integral is far less obvious than it might appear to be.

Aristotle describes politics as the authoritative and comprehensive "master science."[37] This is true in the sense that the "end" of the political science "seems to embrace the ends of the other sciences."[38] Making sense of this claim requires a tremendous expansion of most of our contemporary understandings of politics. The end, or *telos*, of politics is understood by Aristotle to include activities and institutions that we tend to think of as largely independent of politics, including civil society, morality, religion, economy, education, and the family.

While recognizing this implication of Aristotle's definition, however, we must be careful not to overextend it. Although he describes the "end of politics" as "the good for man,"[39] the argument here neither denies nor proposes to eliminate the diversity of realms (the subject matter of other "sciences") into which human activity appears to him to be organized. Politics is inclusive of these other realms in the sense that it may (and should) order and direct them; there is no indication here that Aristotle advances the far stronger argument in favor of erasing all forms of societal differentiation that did exist.[40]

Virtually all of the Greek terms that Aristotle uses to discuss politics derive from the same root. Since this unified vocabulary is not available in English, it is valuable to note that the Greek vocabulary that is translated into such English terms as citizen, regime or constitution, governing class, statesman, and citizenry, as well as those for political science, political office or authority, and political philosophy, are all related to the term for the Greek city-state: the *polis*.[41] As a result, a consideration of the *polis*, as Aristotle understood it, is important to an understanding of his conception of politics, and especially his conception of what is and is not political.

For Aristotle, politics is about what takes place—or should take place—in the *polis*. He identifies and describes the characteristics of a distinctly political community with reference to this particular (from our perspective) species of it. In Aristotle's work, the *polis* is differentiated from entities such as empire or nation, which also existed in his time.[42] One key to this difference is that only the *polis* is guided by a regime or constitution (*politeia*) which, according to Aristotle, makes it possible for the *polis* to promote the goal of human flourishing. This point also allows us to distinguish this conception of the *polis* from the cosmopolitan vision of the Cynics and Stoics. The element of universalism among these thinkers precludes the existence of a particular regime in Aristotle's sense. We must also distinguish Aristotle's *polis* from such modern concepts as the nation-state, or a Weberian notion of the bureaucratic state. As a result, we must be wary of easily ascribing Aristotle's discussion of political community to these non-*polis* forms. Much of this distinctiveness is lost in those English versions of Aristotle's writings that translate the *polis* simply as the "state."

By highlighting the centrality of the *polis* to Aristotle's political thought, however, I do not dismiss the relevance of the latter to other forms of political community, including contemporary ones. Much of what he wrote can offer insight into other forms of political communities if we have an awareness of the particularities of its development. Only by recognizing this particularity of his political thought, however, can any adequate assessment of its broader relevance be made. Surely some of Aristotle's arguments should be recognized as either reflections of, or only relevant to, the particular type of political community (the *polis*) with which he was familiar. The applicability of others, however, may best be seen by recognizing the context in which Aristotle first developed them.[43]

By recognizing that Aristotle's political thought was developed within the conceptual context of the *polis,* we can better appreciate the scope that he envisioned for political activity. We must also consider the character of Aristotelian politics. Aristotle clearly suggests that political rule is distinctive.[44] For one thing, it requires that citizens take turns in ruling and being ruled. Participation in political decision-making is presented as central to the definition of citizenship and to political rule itself.[45] The character of politics, according to Aristotle, is also such that political participation can serve to transform the views of the citizens who engage in it. If citizens "participated in a similar way when joined together as they had when separated," then they would not truly constitute a *polis.*[46]

While this transformation is something that Aristotle presents as an important element of the character of politics, it is not something that he seems to have regarded as intrinsically valuable for all. Clearly, the majority of the population, composed of women, children, slaves, and *metics* (alien residents), was excluded from citizenship and hence from participation. Even for many citizens, moreover, Aristotle argues that the most adequate role that they might play is usually an uninterested or apathetic one that is, as a result of the demands or obligations of their occupations, quite limited.[47]

What I have sketched to this point is an abridged version of Aristotle's conception of politics. Although my discussion of his conception of nature is a bit less abbreviated, it too is not comprehensive. By offering these sketches, I have sought to establish some markers to help define the general territory that these conceptions cover in Aristotle's thought. A more complete explication emerges through a consideration of the relationship between Aristotle's "nature" and his "politics." In the course of this, a more multidimensional picture of both these conceptions will emerge.

Aristotle's Political Naturalism

At the beginning of the *Politics,* Aristotle imagines the historical evolution of human communities. This evolution begins with the household, which is described as the result of what he believes to be natural differences and natural complements between men and women and between slaves and masters. Because this first form of community is, for Aristotle, created as a result of human nature and because the nature of humans is one particular instance of the broader character of natural bodies, he describes it as natural. The end of this household community is the preservation of life for the humans who compose it. The next stage of historical evolution, the village, is characterized as an "extension of the household," composed of extended families or clans, and thus is also natural. Rulership in both forms of community was viewed by Aristotle as of the sort between kings and subjects, an arrangement that he points to as an explanation for the continued existence of kingship within the far larger (and in his sense, nonpolitical) communities of "nations," where the *polis* does not exist.[48]

The *polis*—the distinctly *political* community—is the only form in which self-sufficiency is reached.[49] This claim would make little sense if the concept of self-sufficiency were understood strictly with reference to the material goods necessary for survival. Surely at least the village and nation (if not also the household) could conceivably succeed in meeting this condition. The criterion of self-sufficiency upon which the *polis* emerges as distinctive is with reference to "living well" (for the citizens or potential citizens who are its members; clearly it is not for the slaves, women, and others upon whom the members depend) rather than the less exclusive criterion of merely "living."[50] According to this criterion, Aristotle asserts that even the nation does not qualify as self-sufficient, primarily because he is unable to imagine a regime that could structure such a populace.[51] He suggests that a regime is a condition necessary for humans to flourish in a life well lived and is thus central to the definition of a political community. Like the household, village, and nation, the *polis* is natural, since it emerged from at least the first two of these.

Since "self-sufficiency is an end and what is best,"[52] moreover, and since the central understanding of nature for Aristotle is a teleological one, the *polis* is the most natural community, or the fulfillment of the potential inherent in other natural communities. "For the [*polis*] is their end," Aristotle asserts, "and nature is an end."[53] It is in light of this that Aristotle's contention that "The [*polis*] is thus prior by nature to the household and to each of us" should be understood.[54] For Aristotle, this can only mean that the *polis* is appropriately our end, hence prior *by nature,* although he has already made it clear that it is neither historically prior nor necessary. Maintaining that the *polis* is natural, therefore, certainly depends upon Aristotle's conception of human nature, and the conditions necessary for this to flourish. It also depends upon his teleological conception of nature itself, of which human nature is one manifestation.

The implications of the naturalness of the *polis* are not entirely clear. Because of his emphasis on it at the very beginning of the *Politics*, and perhaps because of a smattering of comments about natural right and justice elsewhere in his works, many have interpreted Aristotle's thought in a way that presumes his politics derive from his understanding of nature and natural teleology. In the remainder of this section I reconstruct the key elements of this interpretation, one that I try to do justice to, even though I find it ultimately unconvincing, as I detail in the following sections.

The first step in viewing Aristotle's political thought as deriving from his conception of nature is the one already described here. For Aristotle, the *polis* is the fulfillment of the potential that he sees within human nature itself. Human nature, moreover, is a manifestation of a more general nature, thus making the *polis* distinctly natural. This has led to a second step, in which many argue that Aristotle regarded a specific type of *polis* as natural: one that is guided by what is naturally right. Given what we already know about the meaning of nature for Aristotle, the identification of the *polis* as natural seems to suggest that the *polis* itself must have an inner principle of change, and that this principle directs it toward its *telos*. When it is made explicitly, this deduction may well seem awkward, both in itself and as an interpretation of Aristotle. After all, it suggests that the *polis* exists as a natural unit apart from its composite parts, i.e., the citizens.[55]

However, the notion that there is a *telos* for the political community seems to be necessary for interpretations that connect Aristotle's political naturalism with a perceived commitment to principles of natural right. This latter linkage is made by many who devote relatively little attention to the question of the teleology of the *polis* per se. Nonetheless, some political teleology must be presumed in order to link the two.

There is another reason the derivative account of Aristotle's theory that I sketch here must rely upon an argument or presumption that the *polis* has a *telos*. This is Aristotle's well-known statement at the beginning of Book One of the *Politics* that "*Every* polis . . . exists by nature."[56]

Without a reliance upon political teleology, this statement would stand in marked contrast to an interpretation by which the truly natural *polis* would be at best exceedingly rare. By its very existence, Aristotle seems to be saying, the *polis* is a manifestation of nature. Such a definition seems to offer little room for nature to be regarded as a higher standard by which the regime of a *polis* could be evaluated.

Overcoming the tension between these points is the work of teleological argument. If the identification of every *polis* as natural means that there is a distinctive natural end for any and every *polis*, then it becomes possible to reconcile two otherwise dichotomous uses of the term "nature" with regard to the *polis*. In the former sense, nature is used as a characterization of every *polis*, while in the latter it is used as a term of distinction for that very uncommon *polis* that has attained its natural end—the *polis* of the best regime.

Leo Strauss maintains that "[n]atural right in its classic form is connected with a teleological view of the universe."[57] Such a formulation offers a concrete "payoff" for Aristotle's political naturalism. It tells us that the premise that the *polis* is natural means that a particular (natural) standard is available to us in our struggle to define the proper basis for organizing and directing it. Because the *polis* is natural, this argument posits, it must have an end that is natural to it. This end appears to be a particular form of the *polis* that is best. Thus, Aristotle's natural teleology appears to point to one regime as the best end of the *polis* precisely because the latter exists by nature. If the *polis* were merely an artificial creation of humans, then nature could offer no standard by which it could be evaluated. Here, nature is contrasted with law or convention (*nomos*). In light of the identification of the *polis* as a natural thing, this contrast has suggested to many that Aristotle regarded nature as offering a standard of what is right or just that is superior to any conventional one.[58]

What initially appears as a description of the *polis* as of a natural type (and hence its location within the realm of nature) is converted into an argument on behalf of nature as a first principle of political organization. Even if we ultimately conclude that this particular conversion is illicit, finding some such "payoff" seems sensible and perhaps inevitable, given Aristotle's insistence upon the naturalness of the *polis*. For this insistence to be significant, it must have some clear consequence for the political theory that follows from it.

How could we know when a *polis* is constituted in a naturally correct manner, according to Aristotle? One answer is that such a *polis* would be structured to promote the "common advantage," rather than the particular "advantage of the rulers." While Aristotle classifies the latter as deviant regimes, he describes the type of regime directed to the common advantage as one that "accords with nature."[59] In this way, it has been said that Aristotle "explicitly extends" a teleological account of nature and natural justice to a theory of regime type.[60] As this interpreter contends,

> For the practical purposes of Aristotle's politics the most important implication of his political naturalism is that the notion of nature provides a criterion by which polises can be evaluated and compared. For a polis may be either in a natural condition . . . or in an unnatural condition. [61]

In this way, the identification of the *polis'* naturalness becomes the basis for an easy progression to Aristotle's discussion of the "best" regime, as presented in the last two books of the *Politics*.

"[T]here is only one constitution that is by nature the best everywhere," Aristotle asserts.[62] While Aristotle's best regime is "constituted on the basis of what one would pray for," it is not intended as an unattainable ideal, as he regarded Plato's *Republic*. "[N]one of these things [preconditions for the best regime] should be impossible," he maintains.[63] This assertion is indispensable if we are to view the regime as natural in Aristotle's sense. Among these preconditions, however, are fairly specific requirements regarding the appropriate size, type of territory, climate, functions, occupations, and character of the *polis*.[64]

This ideal stands in marked contrast to Aristotle's more empirical examination of extant *poleis* in the earlier books of the *Politics*. His earlier examples do not come very close to the combination of preconditions necessary for the best regime. If the *polis* possesses a nature, directed toward the end of living well, then for Aristotle the "best" regime would seem to be the one that allows for the achievement of this end. Such a connection, however, would identify the *polis* as a very unusual sort of natural thing. It would represent the only example of a natural thing that virtually never reaches its *telos*.[65] Yet, as discussed earlier, the notion of tendency or regularity is central to Aristotle's use of the term "nature."

If there is only one particular sort of *polis*, only one regime or constitution that is in accord with nature, then according to Aristotle, it is this regime that is the *telos* of politics. The creation of this one truly natural regime, by definition the best, is the end toward which we should be striving. Only in this *polis* can human flourishing truly be attained. More-over, only in this *polis* would human nature truly be in accord with the rest of the natural order. Humanity would be in its natural place within the cosmos when it is within this political community, and thus here (and only here) the relationship between humanity and the rest of the cosmos would be natural.

An important aspect of this derivative view of Aristotle's political thought is the emphasis that is placed on his account of the "best regime," as developed in the final two books of the *Politics*. This emphasis, or something much like it, is necessary because a derivative interpretation must be able to point to a particular political structure as the distinctively natural one. While many interpreters may then wish to argue that a less-than-best, hence non-natural, regime was often regarded as acceptable in practice by Aristotle, this argument still retains the close connection be-tween nature as a standard and the particular form of the best regime.

This interpretation of Aristotle is central to influential portrayals of him as an originator of a tradition of natural law or, more broadly, natu-ral right within political philosophy. In the absence of such an interpretive framework, the identification of Aristotle with this tradition rests on re-markably shaky grounds. In his entire ethical and political corpus, he devotes only about one page of the *Nicomachean Ethics* to the subject of natural right,[66] and this discussion is remarkably unclear. While it has often been cited as evidence for viewing Aristotle as an advocate of natu-ral law or right in the sense outlined here, there is little in this text itself that provides the basis for this view.[67]

The interpretation has been advanced by and accepted by many thoughtful students of Aristotle because of the combination of two key elements, both discussed earlier. The first is Aristotle's treatment of na-ture, natural order, and teleology. The second is his identification of the *polis* as existing by nature. Once these two elements are in place, a read-ing of Aristotle's highly compressed and ambiguous discussion of natural right is likely to be regarded as support for the conclusion that he envi-sions this as an immutable, higher standard from which a political ideal can be defined and according to which political communities can and ought to be evaluated.

Is this sort of interpretation the one that Aristotle intended to present us with? I am not convinced that it is.[68] Nevertheless, this interpretation does make sense of a number of important ideas within his philosophy. Furthermore, Aristotle offers relatively little in the way of developed, explicit arguments that would challenge this interpretation. There is much that is left unaddressed by this posited relationship between nature and politics, however, and those who regard Aristotle as a dualistic thinker discuss some relevant issues that I consider next.

The Other Side of the Coin: Aristotle as Dualist

The most well-known and influential interpretation of Aristotle that runs counter to the derivative interpretation is presented by Hannah Arendt. According to Arendt, Aristotle's political thought is distinguished by his identification of the Greek *polis* as the "sphere of freedom."[69] In Arendt's reading of Aristotle, freedom in the *polis* is explicitly contrasted with nature. Nature here is equated (and conflated) with the realm of necessity: the household (*oikos*) in which participants are occupied with the production and reproduction of all that is necessary to sustain life itself. "Natural community in the household therefore was born of necessity," Arendt contends, "and necessity ruled over all activities performed in it."[70] As a result, both the primary participants in this realm—women, slaves, children—*and* the fundamental concerns of this realm—economy, agriculture, parenting, gender relations, slavery—are said to be excluded from public, political life. Thus Arendt also contends—citing a pseudo-Aristotelian text—that

according to ancient thought on these matters, the very term "political economy" would have been a contradiction in terms: whatever was "economic," related to the life of the individual and the survival of the species, was a nonpolitical, household affair by definition.[71]

Wendy Brown (critical though she is of Arendt's normative perspective) summarizes the point well when she concludes that "Greek men, as Arendt tells the story, . . . suppressed and violated their connectedness with others and with nature."[72]

Whatever we make of this argument on its own terms, the most relevant characteristic is that this Arendtian view is presented as Aristotle's.[73] From this point of view, a dualism between politics (identified with freedom) and nature (identified with necessity) is not a distinctly "modern" creation, as many who see such a dualism in Hobbes's thought contend. A clear expression of it can be traced back at least as far as Aristotle and the ancient Greeks. Indeed, for Arendt (although *not* for environmentalist and feminist theorists who accept her characterization of the Greeks), the tragedy of modernity is rooted in the demise of this dualism, as a result of what she terms "the rise of the social."[74] For Arendt and many other political theorists however, identifying past philosophers as adherents of a dualistic view was not intended as an accusation. Arendt celebrates the nature-denying freedom that she identifies with her description of the Aristotelian *polis*. For environmentalists, feminists, and others, however, this interpretation can be both explanatory and critical.[75]

What basis do Arendt and others have for asserting the existence of a dualism between nature and politics in Aristotle's writing? Above all, Arendt emphasizes Aristotle's embrace of the division between the *polis* as the realm of public life and the *oikos* or household as the realm of private life.[76] Aristotle clearly suggests such a division when he differentiates between those "things . . . which are present in cities of necessity" and the "parts of a city."[77] The former, "things" present of necessity, include all the participants in the maintenance of the household, including women and slaves. Aristotle makes no effort here to differentiate between human and nonhuman, or between animate and inanimate "things" or "possessions" in this context.[78] By contrast, "parts," for Aristotle, refers only to the citizens of the *polis* itself. Aristotle clearly does not imagine these "parts" to be bound to the *oikos* in the sense of spending their days occupied with farming, craftmaking, parenting, or other activities central to this realm.[79]

The assertion that "free" citizens are not bound to the realm of private necessity—the *oikos*—is a relatively uncontroversial claim to make about Aristotle's view. Most interpreters who do not embrace the dualist interpretation of Aristotle's thought would nonetheless acknowledge that he explicitly excludes those humans most closely tied to the *oikos* (women, slaves, children) from participation as citizens in the *polis* and that he sees a sufficient amount of leisure as a precondition for the life of a citizen. Establishing a dualist interpretation of Aristotle's thought, however, requires more than this.

Whereas the inhabitants of the *oikos* were occupied with the necessities of survival and life, according to Arendt's interpretation of Aristotle, the citizens of the *polis* were focused on something "of an altogether different quality;" in Aristotle's words, the "good life" (or "living well"). According to Arendt, life in the *polis:*

was "good" to the extent that by having mastered the necessities of sheer life, by being freed from labor and work, and by overcoming the innate urge of all living creatures for their own survival, *it was no longer bound to the biological life process.*[80]

Here, Arendt presents Aristotle as not merely differentiating "freedom" and participation in the *polis* from day-to-day, full-time responsibilities in the *oikos,* but is also equating the latter with the "biological life process" itself. Hence, the *polis* appears distinguished by its *anti*naturalness—its opposition to biology and life—just as the *oikos* is equated with the fulfillment of these necessary and natural processes. Here, not only are the participants in the *polis* differentiated from those of the *oikos,* but the concerns of the former are also distinguished by their exclusion of all questions connected with biology, survival, life, labor, . . . *nature.* It is this crucial equation of nature with the realm of human activity most closely interactive with the physical world (necessity) that is at the heart of Arendt's influential interpretation of Aristotle's philosophy. Authors such as Murray Bookchin, Val Plumwood, Victor Ferkiss, and Wendy Brown who also identify Aristotle with a nature–politics dualism present their interpretations of Aristotle largely as filtered through Arendt.[81]

By connecting nature with the realm of necessity, the Arendtian interpretation allows us to conclude that Aristotle *intentionally* proposed a dualism between nature and politics. One need not, however, begin with this connection in order to conclude that there is a dualism between nature and politics in Aristotle's thought. David Keyt, reading Aristotle analytically, argues that Aristotle created a dualism for quite different reasons, and quite unintentionally. Keyt maintains that

according to Aristotle's own principles the political community is an artifact of practical reason, not a product of nature, and that, consequently, there is a blunder at the very root of Aristotle's political philosophy.[82]

What is the basis for this claim, which strikes at the heart of Aristotle's stature as a political philosopher? It is that interpretations that even remotely resemble the derivative account of the nature–politics relationship, in which Aristotle's political naturalism is emphasized, fail to make adequate sense of quotations such as the following from Book One of the *Politics:*

Accordingly, there is in everyone by nature an impulse toward this sort of partnership [the *polis*]. And yet the one who first constituted [a *polis*] is responsible for the greatest of goods.[83]

The first sentence here seems to reinforce Aristotle's emphasis on the naturalness of political community and certainly reiterates his contention that man is a "political animal," claims that should be familiar by this point. The second, however, refers to a lawgiver or statesman as one who "constituted" the *polis,* and is thus "responsible" for its goodness. In this sentence Aristotle clearly points to the *polis* as an artifact of human creation. The challenge that this assertion poses for a derivative account of Aristotelian politics should be evident. If the regime that constitutes the *polis* is itself artifice, then there is no basis for identifying a particular form of this artifice as naturally best. The seeming contradiction between these two claims is the basis for Keyt's argument that Aristotle "blundered."

Some commentators have sought to gloss over this difficulty by suggesting, for example, that "art co-operates with nature" in Aristotle's understanding of the development of the *polis.*[84] This is a gloss because it fails to regard Aristotle's conception of nature as meaningful. If we were to accept the contention that art can cooperate with nature, while maintaining the naturalness of a given phenomenon, then essentially everything would be natural, and nature would no longer serve Aristotle as a term of distinction. As Keyt points out, by this definition not only *poleis* but also houses, paintings, and all other human creations would properly be understood as natural for Aristotle, since all are the result of "art" in cooperation with natural materials. In truth, "no artifact comes into being without the aid of nature," and so Aristotle's claim that the *polis* is natural appears to be reduced to an inconsequential truism. In this interpretation, there would be nothing distinctive about Aristotle's claim that the *polis* was natural. As Keyt explains,

The difference between . . . art aiding nature and nature aiding art, is that natural conditions and natural events such as health and parturition are ends that nature can, and indeed usually does, attain unaided (see *Phys.* II.8.199b15–18, 24–26)

whereas objects of art such as paintings, statues, and poems are never produced by nature alone.[85]

Rather than claiming that we can derive a distinctive model of the *polis* from its proclaimed naturalness and utilize the art of a lawgiver or statesman in a limited manner to achieve nature's end, Keyt's Aristotle has only truly established a natural basis for the *polis* in the sense that marble provides the natural basis for the sculptor's creation of a statue. This sense, Keyt clearly presumes, is inconsequential.

Keyt concludes that Aristotelian politics is really artificial rather than natural. The centrality of a lawmaker in Aristotle's account of the creation of the *polis* makes it inextricably artificial, hence unsusceptible to claims to have identified distinctively natural standards of right to be held up as measurement. The derivative account is undermined by the role that Keyt identifies for artifice in the creation of the *polis* and as a result he finds himself forced to conclude that Aristotle must have meant to offer a derivative account, but inexplicably "blundered." Nature cannot offer us the answers to political dilemmas that Aristotle seems to have suggested it could, and so, Keyt argues, we must reject the claim that he offers us a way of linking nature and politics. Once again, a duality between nature and politics becomes central, although for different reasons than Arendt had postulated.

In the previous section of this chapter, I outlined the key elements of a derivative interpretation of Aristotle's writings on nature and politics. Clearly, there is much in his writing that might encourage us to embrace some version of this interpretation, and many have done so. Yet what may have been clear earlier should certainly be so now: The derivative interpretation cannot neatly integrate all the elements in Aristotle's political philosophy. The dualist interpretation reveals some of these unintegrated elements: Aristotle's exclusion of the members of the *oikos* from participation in the *polis* and his emphasis on the centrality of the creative role of individuals in the formation of the *polis*. These elements have led the theorists discussed in this section to offer a contrasting interpretation in which Aristotelian politics is distinguished by its clear separation from something deemed "nature," and consequently from any standard of natural right.

It is striking, however, that dualist interpreters make relatively little effort to explain how they could integrate the quite significant elements of Aristotle's philosophy that are central to the derivative view. For Keyt, the claim that Aristotle has "blundered" absolves him of the need to elaborate a consistent Aristotelian argument. After all, inconsistency is the central problem that leads Keyt to conclude that Aristotle committed a blunder in the first place.

For Arendt, and those who follow her influential line of interpretation, the omission is more problematic. Claims that "man is by nature a political animal" and that the *polis* "exists by nature" are among the most familiar and oft-repeated lines of Aristotle's political philosophy. To do as Arendt does, advancing an interpretation in which Aristotle is said to have erected an insuperable barrier between nature and politics without ever even acknowledging these contrary claims raises serious questions about the coherence of this dualist interpretation. Despite these questions, however, the dualist interpretation succeeds in presenting us with a starkly delineated image of Aristotelian politics that brings into focus some very real and significant elements in his writings. In the following section, I navigate a route between the derivative and the dualist extremes in a way that seeks to respect the significant, although limited, insights of each approach.

Nature and Politics in a Dialectical Relationship

The dualist interpretations of Aristotle do reveal weaknesses in the derivative account. As I have argued, however, they do not offer a wholly convincing or comprehensive account of Aristotle's discussion of the nature–politics relationship. Keyt, who is explicit about his inability to offer a coherent, comprehensive account of Aristotle's thought, turns the tables and blames Aristotle himself for this failing. Perhaps this is the best we can do; certainly any effort to formulate an airtight interpretation of Aristotle's thinking on this topic is likely to be weakened by the disparate elements of his own writing. Nonetheless, in this section I sketch an alternative interpretation of Aristotle's discussions of the subject at hand, one that sheds some light on the dilemmas of the nature–politics relationship.

Although Keyt himself points to a dualism between nature and politics in Aristotle's thought, his accusation that Aristotle blundered is based on a presumption that consistency would have required the *polis* to have been derived from a prior conception of nature. In this view, Aristotle's claim that the *polis* is natural should lead to the conclusion that his "best" regime is the *polis'* natural end, although in fact it does not.[86] Thus Keyt accepts the derivative interpretation of Aristotle's intentions, even while accusing him of failing to achieve these intentions. Keyt's dualism, in other words, is premised upon a particular—derivative—understanding of what it would mean for nature and politics to be related.

It may, however, be possible to sketch an alternative understanding of the relationship between nature and politics in Aristotle's thought that does not rely upon deriving the *polis* from nature, and hence is not invalidated by Aristotle's own observation that the establishment of the *polis* is an act of artifice.

A preliminary task necessary for explicating a nonderivative relationship between nature and politics is to sort out the relationship between the *polis,* the *oikos,* and other realms of human activity in Aristotle's thought. To begin a comprehensible discussion of the *polis* in relation to these other realms requires, in turn, that we clarify the meanings that Aristotle attributes to the *polis* itself.

R.G. Mulgan has pointed out that Aristotle uses the term *polis* in two different senses. He calls these the "inclusive" and "exclusive" senses.[87] While this is not a distinction that is clearly made by Aristotle, it is one that can be seen throughout his political writings:

On the one hand, *polis* refers exclusively to one particular aspect of the city-state, the specifically political institutions in contrast to other institutions and groups, such as families, business partnerships, and so on; it is the political community as distinct from all other communities. On the other hand, *polis* may mean the whole society, including all these other communities.[88]

It is the "exclusive" definition that more closely approximates our own everyday understandings of politics and government. While we may disagree about the proper size or range of functions for government (and while we may certainly disagree with Aristotle on these questions), we would rarely confuse this everyday sense with a conception of politics that includes the whole social order. This exclusive definition of the *polis* also corresponds to, and sheds light on, Aristotle's discussion of regimes or constitutions (*politeia*).

Aristotle's famous sixfold classification of regime types centers on the composition and character of the participants in the governing institutions of the *polis*. Aristotle's three "correct" regimes—kingship, aristocracy, and polity—are differentiated by the segment of the population involved in governing, and hence the segment that are citizens.[89] Each of his three "deviant" regimes corresponds to one of these "correct" ones, differing only in that the character of participation is focused here on the self-interest of the rulers.[90] Even where Aristotle applies these neat ideal types to the messy world of actual regimes, the focus remains on governing institutions in particular, not on the broader "inclusive" understanding of the *polis*.

It is also this narrower "exclusive" sense of the *polis* that Arendt and others describe as the sphere of freedom, in contrast to the *oikos*. The "inclusive" *polis* could not be contrasted with the *oikos,* since the latter is within it. Consider decision-making in the *polis.* As we have seen, Aristotelian politics characteristically involves interaction and power sharing among roughly equal individuals, obliges them to serve as judges, and requires that they be capable of both ruling and being ruled.[91] All this reinforces Aristotle's reliance upon leisure and freedom as requirements of citizens in the *polis*[92]; to be "part" of the *polis,* rather than a mere "necessity" within it.[93]

The preconditions of citizenship and participation are explicitly identified with freedoms defined in opposition to the realm of the *oikos.* Moreover, Aristotle goes beyond observing an empirical distinction in the Greek *polis* between the freedom of citizens and the absorption of women, slaves, and others in the realm of necessity. He attempts to offer a rationale for such a division, defining the nature of women and (natural) slaves in such a way that their devotion to the *oikos* and exclusion from politics appears justified as the fulfillment of their nature.[94] Aristotle also advances normative claims, in the case of slavery "according to law,"[95] that may offer a basis for a social critique of which individuals are assigned to the contrasting realms of "necessary" and "free" forms of activity. Nonetheless, these claims reinforce the existence of such a division and the expectation that sufficient and appropriate individuals or groups can be identified to populate each side of the divide.

Participation in the *polis* thus depends, for Aristotle, upon a freedom defined in opposition to necessity. The realm of necessity, the *oikos,* moreover, is clearly the one that is closely tied to interactions with the physical world. The form of political activity is in clear contrast to the form of activity in the *oikos,* and Aristotle excludes those devoted to the latter from participation in the former. All this, we have already seen, is addressed by dualist interpreters. Indeed, if we could limit the definition of "dualism," as Val Plumwood does at one point, to "a relation of separation and domination inscribed and naturalised in culture," then the dualist interpretation of Aristotle would be largely accurate and compelling.[96]

We can see that in the *Politics* Aristotle certainly does use the language of nature to legitimate the ordering of human spheres and groups, and in particular to legitimate the relationship of *oikos* to *polis* and the membership of each. While it may well be that he is accepting the status quo of his own society with regard to gender and slave relations, this acceptance, in combination with the use of his normative language of teleological nature, establishes an image of a fixed and legitimate *natural ordering* of humans and spheres of human activities that is evidently divisive and inegalitarian.

Although Aristotle uses the language of teleological nature to legitimate his supposedly natural order of human society, this is not consistent with other elements of his philosophy, including what I take to be among his most powerful and insightful arguments about politics. This becomes clearer when we shift our attention away from the exclusive definition of Aristotle's *polis* and toward the inclusive one. It is when he uses the inclusive sense that Aristotle speaks most convincingly of the *polis* as natural and of political science as the "master science."

With regard to this science, everything in the *polis* is a subject of interest and is also *potentially* political in the (exclusive) sense of being a subject of active involvement by the governing regime. The fact that Aristotle regards the *polis*, in the inclusive sense, as natural tells us that he regards it as a natural end for humans to form associations, structured by a regime, in order to live self-sufficiently and well. This is the conclusion that can be drawn clearly from the origins given in the beginning of Book One of the *Politics*, which describes the *polis* as having emerged from the village, which itself is the result of the aggregation of an earlier natural form of human association—the household.

These are bold and explicit assertions: that both the *polis* and the *oikos* are natural, that the latter is properly incorporated within the former, and that politics is the appropriate realm for determining the ends and objectives of the household. Here, politics is understood as the conscious consideration and possible delineation of the ends of technological development, economic and market activity, and private and household relations. Here, the human activities most actively intertwined with the physical environment are stripped of any illusion of necessity. The contrast with Aristotle's supposedly natural ordering of human society should be clear. The activities of the *oikos* become a fundamental component of political decision-making and thus can be limited and directed by the political community's own understanding of the appropriate ends and purposes of these activities.

As such, Aristotle's argument for the recognition of the embeddedness of the *oikos* in the *polis* resonates not only with many contemporary environmentalist arguments,[97] but also with the quest of socialists to politicize economic activity, and with feminists who assert the "personal as political."[98] In this way, Aristotle suggests to us the possibility for *politicizing* precisely those questions and relationships that his discussions of slavery, the role of women, and the division between the *polis* and *oikos* appeared to have "naturalized."

By describing the *oikos* as subject to politics, Aristotle allows for questioning or restructuring of the relationships of the *oikos* itself, as well as the boundaries between it and the realm in which political deliberations take place. This, of course, is precisely the intent of contemporary feminists who argue that "the personal is political."[99] That feminists would have something in common with the misogynist Aristotle on this point is ironic. Aristotle can advance this argument precisely because he does not seem to anticipate a challenge to the formal boundaries that he posits between the realms of *oikos* and *polis*. These boundaries are presented as natural in that they reflect his own understanding of the natural ends of those who populate each realm.

Thus, on the one hand, Aristotle offers an encompassing scope for political deliberations, one that allows for the direction of all household activities within it. On the other hand, by appealing to nature to justify the boundary between the *oikos* and *polis,* he acts to prevent any political deliberations about this boundary itself. In his own time, this latter step may have appeared relatively noncontroversial to his readers. From our perspective, however, we can argue that by politicizing the relations of the *oikos,* Aristotle cannot prevent evaluations of the naturalistic claims he used to assign women and slaves to roles exclusively in the private sphere.

To suggest otherwise, as Arendt does, is finally to imagine an ethereal politics without real power or even any external referents. This is not Aristotle's politics. Aristotle presents issues of material necessity (such as the food supply) as central to political leadership and deliberation.[100] Moreover, his discussion of the "best regime" in Book Seven of the *Politics* makes it clear that a plethora of questions related to geography, land use, population density, and natural resources are central to political arguments and regimes, whether or not we are hoping for the best possible one. Aristotelian politics is by no means strictly self-referential, but only by distinguishing the differing meanings that Aristotle attributes to both the *polis* and to nature, as I have done here, can we appreciate why this is so. Politics channels and directs human activities, including the many ways that these bear upon the physical world. Nature cannot be excluded from political deliberations.

Aristotle argues that it is with respect to the regime—the structure and institutions of governance—that the (inclusive) identity of the *polis* is created and maintained.[101] This is an important and realistic recognition of the fact that the governing institutions of the *polis* do in fact affect and structure all the other activities and realms within it. This emphasis on the regime as providing an inescapably important context for other human activities within the *polis* is a distinguishing characteristic of Aristotle's ethical and political thought. It is also one that offers a clear contrast to many liberal theories of politics.[102] Again, however, this recognition ought not lead us to conclude that this influence presumes the nonexistence of other activities and realms distinct from the institutions of governance;[103] in other words, it should not be conflated with a defense of totalitarianism.

Aristotle arrives at a conception of political activity that requires a commitment of time and energy—in other words, freedom or leisure—in order to fill the role of citizen adequately. How this freedom is to be achieved and to whom it is to be given, however, are not questions that preoccupy him. Instead, he accepts the conventional answers to these questions within the Greek *polis* of his day. More pointedly, he goes some way toward "naturalizing" these political answers through his assertions about the inferiority of women and the existence of natural slavery. This position is significant, because only if he can maintain the formal boundary between those who participate in political activity and those who are restricted to the household *as natural* can Aristotle exclude an examination of these boundaries from his encompassing characterization of the *polis*.

Today, most of us have little trouble recognizing that Aristotle's ascription of limitations to women and ("natural") slaves that would render them unfit for participation in the *polis* is anything but natural. And there is nothing about a teleological view of nature that would prevent us from arriving at this same conclusion from within it.[104] Aristotle's boundaries and exclusions are political in an inclusive sense—not natural.

At the same time, his argument allows us to recognize that politics can and does play a key role in shaping and defining our relationship with the rest of the natural world defined as the collection of all things that have a nature. From our perspective, we can identify Aristotle's politics as highly authoritarian in that political decisions are made by a minority composed of free male citizens and are imposed upon all others. This authoritarianism may turn out to be more severe because the decisions of this minority are able to reach into all aspects of life, our relationship with each other, and the rest of the cosmos. While this political arrangement is not natural, it plays a central role in defining our relationship to nature.

The assertion that we must understand the nature of our world in order to know how to properly live and act in it finds support within Aristotle's philosophy. And yet, Aristotle's natural teleology cannot provide him with principles of natural right to direct political action. What, then, does his assertion that the *polis* is natural tell us? For one thing, it tells us that Aristotle believes that the *polis*—as well as those who act in it—is located in the natural world, and hence that the *polis* is formed in accordance with the same natural principles as all other things. Moreover, the political judgments made by citizens within the *polis* are appropriate determiners of the ends of all other and historically earlier forms of human community—most especially the household or *oikos*. It is in this latter sphere that Aristotle places the production and reproduction of all that is necessary for human life. Farming, technics, economic exchange, childbearing and rearing, and labor are among the things vital to human existence that he assigns to the realm of the household.

Among those, such as Hannah Arendt and Wendy Brown, who claim Aristotle's politics are artificial, this close association between *oikos* and the physical world suggests a sphere distinguished by its naturalism. Since Aristotle also at times contrasts this sphere with the public world of the *polis,* it becomes easy to presume that the latter is characterized by its non-naturalness. Yet Aristotle maintains otherwise at the very beginning of his *Politics.*

Unlike these recent interpreters, he does not identify the realm of necessity as a *distinctively natural* one. Indeed, he could not identify it in such a manner, since his justification for both the *polis* itself and the identification of those appropriate to participate in it is also based on criteria that he describes as natural. The *polis* "exists by nature,"[105] and, according to Aristotle, participation in it is the natural end of those with a human nature that enables them to engage in the activities of citizenship.[106] Aristotle proposes a hierarchy of realms in which the *polis,* understood exclusively, is elevated above the *oikos,* but both reflect a single understanding of nature and hence are natural in the same sense.

Conclusion

By utilizing an inclusive/exclusive distinction with regard to Aristotle's discussion of the *polis,* we can say that it is in the former sense that he writes of the naturalness of the *polis,* while in the latter sense any particular regime is necessarily a form of artifice, hence is lacking a natural *telos.* Aristotle does try to say more than this; for example, describing who should be included as a part of the *polis,* but this cannot convincingly be identified as natural, either from our perspective or from within his own broader argument.

Understood in this way, Aristotle's nature is able to "tell" us that living together in a regime is a natural end because it is necessary to promote human flourishing. It is unable to provide answers to questions about the particular character and scope of this (political) regime, or its relationship to other realms or associations within the *polis.* What sorts of economic relationships, household structures and roles, technologies, or educational forms ought to be fostered within the *polis*? These are the questions that Aristotle suggests ought to be answered with an eye to promoting the natural end of human well-being.

Given the heterogeneity of his conception of nature, this natural end urges us to pay attention to the particularities of place in the formulation of answers to such questions. As a consequence, it should not be surprising that Aristotle often focuses his political analysis on a set of concerns that are highly relevant to the ways in which nature and politics interact. Issues of geography, resource and land-use planning, and population size are among those elements central to the formation of the *polis*.[107] The connection between the conditions for "living" and those for "living well" is an intimate and continuing concern for Aristotle. The answers to particular questions about these issues, however, will be developed within the regimes of particular *poleis*. They will be the result of deliberations that occur within artificial structures that shape their outcome and thus are contingent upon these structures. Nature and human nature provide the inescapable context within which discussions about the boundaries between various realms of practice take place. The delineation of these boundaries, however, is never truly natural. The boundaries will also reflect the regime that constitutes the process by which their delineation takes place.

If, in an effort to overcome the artificial role of the regime in this process we sought to eliminate its role, then we would be essentially collapsing this exclusive sense of the *polis* into the inclusive sense of it. Yet, for Aristotle, the existence of some regime is a necessary element in the original definition of (this inclusive sense of) the *polis* itself. As a result, it seems that although it is natural to form and live within the *polis*, Aristotle cannot maintain that either the structure of or the decisions within the *polis* can be evaluated by their naturalness. Politics is indeed natural for Aristotle, but political answers are not natural ones.

III

Political Nature

6

Nature, Politics, and the Experience of Place

Here I discuss some of the common lessons learned from my examinations of nature, politics, and the relationship between them. As we have seen, Aristotle and Hobbes are prominent figures in the arguments for both a dualist and a derivative interpretation of the nature–politics relationship. Moreover, a reading of their writings makes it clear that these interpretations are not completely fanciful and do have a basis there. Yet nature is not radically separated or bifurcated from normative political thought in the West, nor is this political thought derivative of a particular conception of nature. To the extent that my reading of these two prominent philosophers challenges the dualist and derivative view, then I will also have challenged the broader claim that Western thought as a whole is best characterized in this manner. After all, a claim about the character of the nature–politics relationship in the history of Western thought that has to cite both Hobbes and Aristotle as exceptions to the rule would be unconvincing from the start. Of course, no study of just Hobbes and Aristotle—significant though they are—could suffice to establish the conclusion that *no* thinkers in Western thought have consistently advanced either a dualist or derivative view. This is not a claim, however, that I seek to advance.

Clearly, there are great differences between Hobbes's and Aristotle's conceptions of nature, on the one hand, and their conceptions of politics, on the other. I have detailed many of these differences in previous chapters and have also noted Hobbes's own vehement rejection of Aristotle on these points. Nonetheless, in this chapter I wish to discuss a singular set of arguments that have emerged through the consideration of the relationship between nature and politics within the works of each philosopher. It may be surprising to find commonality on such a central theme in the works of these two disparate thinkers.

In order to be clear about the arguments that follow, then, I wish to emphasize again that I am not arguing that Hobbes or Aristotle intended to present the relationship between nature and politics in the way that I discuss, at least in any clear manner. Both did, however, offer exceptionally thorough treatments of their understandings of nature and politics. Their texts thus provide an uncommon opportunity to explore this relationship in a way that moves beyond the tendentious and ultimately less interesting question of intention. The relationship that emerges from this exploration is neither derivative nor dualistic. That thinkers otherwise so different can both lead us to this conclusion should serve to reinforce its validity.

I have sought to describe some of the persistent dilemmas that emerge in any attempt to get their arguments to fit into the derivative or dualist frameworks. These dilemmas have not gone entirely unnoticed by other interpreters, but many have dismissed them by arguing that Aristotle and Hobbes made admirable but ultimately inadequate efforts to fit one or the other of these frameworks.[1] I regard this as an unsatisfying conclusion. Instead, the dilemmas themselves can point us toward a more adequate, although less analytically precise, alternative framework. My goal in this chapter is to sketch this framework. To do so, it will be helpful to review and draw together those elements of the dualist and derivative views that *are* convincing. I then use this composite to shed light upon dilemmas central to my previous chapters.

Lessons from the Derivative Interpretation

First, derivative interpreters have claimed that nature can offer us a principle, often presented in the form of a conception of "natural law" or "natural right," that can provide the direction needed for the political ordering of human society. Although it is often arguments about *human* nature that receive the most attention in these formulations, we have seen that these arguments rely, either implicitly or explicitly, upon a broader conception of nature to establish the truth and the invariability of their vision of humans. Not only can a conception of nature provide such direction, but many have pointed to Aristotle and Hobbes as preeminent and successful examples of a persistent effort within Western thought to do so.

From this point of view, then, the political order advocated by a theorist will be compelling only if the underlying conception of nature that serves as its source of direction is true. From within this framework, then, political debate begins and largely ends with an attempt to establish or reject the truth of a particular conception of nature itself. If a derivative interpreter appears to reject the history of Western political thought, this can only result from the conviction that this thought is derived from incorrect views of nature.

The very premise of this interpretation, however, is one that I have been challenging. Rather than rejecting or embracing the particular conception of nature offered by Hobbes or Aristotle, I have focused on the premise that it could serve as a source of political direction. If we come to see that it cannot—that whatever their intentions, neither Hobbes nor Aristotle were able to establish a relationship in which politics is derived from a conception of nature—then we are compelled to shift the focus of debate. While I have argued that this premise is wrongheaded, it certainly is not absurd. Aristotle clearly does proclaim the *polis* to be natural and proclaims man to be a *polis* animal. Hobbes does present his natural philosophy as a necessary premise for his political philosophy. In both cases it is evident that the realm of politics is situated within a broader category that each theorist understands to be natural. In this sense, both philosophers understand humanity itself to be a part of nature. A political order that ignores this nature or fails to adequately understand its significance can be said to be deficient. This understanding lies at the heart of the otherwise very different philosophies of Aristotle and Hobbes. It is rooted in the link that each elaborates between nature in general and human nature in particular. A link at this point is one that cannot be dismissed.

To locate humanity within an encompassing conception of nature demands that we recognize ourselves as natural beings. Understood in this way, nature *constitutes* who and what we are. Aristotle and Hobbes share this conviction with contemporary environmentalists: We cannot properly understand humanity without recognizing its embeddedness in nature itself. In this view, if we want to know how to live and act properly in the world, we must first understand the nature of the world itself. The degree to which the quest to understand nature is central to the work of both Aristotle and Hobbes is difficult to overestimate.

Conversely, all human action affects the nature within which we are embedded. Whatever else humans are, we are inescapably material beings whose existence and actions are embedded within the world around us. Whether we seek to "live lightly upon the earth," or engage in a Baconian quest for power over nature, we inevitably shape and alter our physical surroundings and other beings, including, of course, other humans.

We need not conclude, however, that all the ways in which humans act in nature can be directed by a proper understanding of that nature. This is the promise of the derivative interpretation, but it is one that relegates politics to a limited and virtually invisible role as the executor of natural directives. Neither Aristotle nor Hobbes is a convincing candidate as a defender of this role for politics.

We have seen that the Aristotelian *polis* and the Hobbesian sovereign are both distinguished by the inclusiveness of their subject matter. Politics is the "master science" for Aristotle; the Hobbesian ruler is presented as sovereign over—and interpreter of—*all* realms. All human activity is embraced within these political conceptions. As a consequence, it is more appropriate and convincing to argue that nature is the inescapable *subject* of politics so conceived. We can conclude that nature constitutes politics for Aristotle and Hobbes because we have seen that nature is constitutive of humanity and that politics is understood here as inclusive of all forms of human activity.

Although the constitutive understanding of nature begins at the same point as the derivative interpretation, they lead in different directions. For the former, politics allows us to develop answers to the multitude of questions and challenges posed by our condition as beings embedded within nature. For the latter, this nature appears to offer answers for the human condition itself. While I wish to highlight the argument that nature is constitutive of humanity and humanity's politics, I also wish to emphasize the gulf between this argument and one that asserts a derivative relationship between nature and politics.

The human interaction with the rest of the natural world is inescapably central to the decisions made within the polity and ought to be recognized as such. It may seem commonsensical to recognize that nature is constitutive of human societies and polities; it certainly is so to most environmentalist writers. However, the consequences of this observation are not insignificant. It enables us to see that our relationship to the natural world is not external to politics, and is not merely instrumental, but constitutive of who we are.[2] In a sense, this is an empirical conclusion. Politics is at least potentially about a whole range of issues—war and peace, economic activity, technological development, culture, gender roles, housing, transportation—all of which affect our interaction with the world around us. The consequence is that we should recognize the ways that politics shapes our relationship with the natural world and hence the shape of that world itself.

A recognition of nature as constitutive of politics can prompt us to see that addressing the challenges posed by our relationship to the physical world is central to what politics is. Political actors, whether sovereigns or citizens, ought not to behave as though the natural world is merely an insignificant or unchanging backdrop for their actions and decisions. Instead, whether directly or indirectly, political decisions affect and shape the natural world in the most profound and consequential ways. It is important to note here that recognizing and addressing these challenges and questions does not ensure that they will be answered in a way commensurate with environmentalist hopes, a point that I discuss in the next chapter. While it is not sufficient, this recognition is necessary[3] and could ensure that these challenges are reviewed as central to all forms of political action or decision-making.

A regard for nature as constitutive of politics precludes the dualist interpreters' contention that Western thought has established a firewall between nature and politics. It in no way, however, certifies Hobbes, Aristotle, or Western thought as a whole as "environmentalist" in any meaningful contemporary sense. This is a point worth emphasizing. It suggests not only the conceptual possibility, but also the feasibility of a theoretical position that is simultaneously antidualist and antienvironmentalist. This is significant because it suggests that *even if* dualism were dominant throughout Western thought, overcoming it would not ensure the realization of environmentalist goals. Among those who recognize nature as constitutive of humanity, a wide range in attitudes and stances toward the natural world can remain.

So what value is there in emphasizing nature as constitutive of politics? It allows us to place questions related to nature at the center of our political agenda. Moreover, it allows us to recognize that all questions on the political agenda are related to nature, directly or indirectly. To view this emphasis as offering a source of political authority, however, is to fail to see that politics itself is a realm in which differing interpretations are central to disagreement. If we begin to see politics in this light, then we can also see it as a realm in which disagreements rooted in differing interpretations of the natural world and the appropriate human stance toward this world can be brought to the foreground. All understandings of our relationship with nature are ultimately political in this sense, although the question of whether any government actions should be undertaken is one that is left unanswered.

Dualist Insights

The argument on behalf of nature as constitutive of politics by definition precludes a dualism. Yet this is not the whole story. To fully appreciate the relationship between nature and politics, we must also address those elements and arguments that are central to the dualist interpretation, even though this interpretation is not persuasive as a whole.

Hobbes *does* present the commonwealth in contrast to his state of nature. Aristotle *does* elevate the deliberations of the *polis* above the realm of the *oikos,* in which the necessities of life are obtained and fashioned in close association with the nonhuman natural world. In both cases, an explicit boundary is drawn. On the one side of this boundary lies an avowedly natural condition or set of relationships—in Hobbes's state of nature or in Aristotle's household. On the other side of the boundary lies a condition characterized by something that appears, to dualist interpreters at least, as different in kind from this characterization of nature. It is the space on this side of the boundary that Hobbes describes as the commonwealth; a sphere said to be *artificially* ordered by the sovereign. It is on this side of the boundary, as well, that Aristotle locates the public, deliberative life of the *polis* members, for which freedom is said to be both a necessary prerequisite and a defining quality.

Nonetheless, we cannot conclude that a dualism between nature and politics exists in these works. Politics is an inescapably human endeavor and humanity, we have seen, is constituted by an inclusive understanding of nature itself. If a true dualism between nature and *humanity* is not possible within the theoretical world of either Aristotle or Hobbes, then a dualism between nature and *politics* is equally impossible. Instead, what these theorists seem compelled to do is to *differentiate* realms in which human activities take on distinct roles. This act of differentiation can be seen as a process of mapping boundaries between particular understandings of political deliberation, power, productive and reproductive activities, gender roles, and civil society. In themselves, these acts of differentiation and mapping are both inescapable and unproblematic. A theory that avoided such differentiation would be incomprehensible. It is hard even to conceive of what it would mean for human life and activities to exist as an undifferentiated whole. Of course, any actual act of differentiation is bound to raise doubts and disagreements. By recognizing these acts for the human constructions that they are, however, we can view them as both profoundly political and independent of any true separation from nature.

By delineating this boundary, both Hobbes and Aristotle emphasize that political decisions are not and cannot be appropriately understood as *distinctively* natural ones. Political actions are acts of freedom, according to Aristotle. Sovereign power is a consequence of what Hobbes calls artifice. In neither case is political authority derived from a natural directive. While these distinctions do not represent a true cleavage or dualism between nature and politics, the existence of a boundary is nonetheless both evident and significant. In order to establish their conception of politics, Hobbes and Aristotle rely upon a realm legitimated by its apparent "naturalness," yet differentiated from the whole of nature. Environmentalists, too, often rely upon such a differentiated realm in order to advance their normative arguments. Yet we have seen that the protection of this realm is often viewed as necessitating an "ecological worldview," which also appears legitimated by its "naturalness." Yet environmentalists are no more able to establish a conception of politics on this basis than Hobbes or Aristotle.

Defining the realm of "politics" as well as "nature," "economy," "civil society," and "private life" is a distinctly human act. As an interpretive act, it is laden with theoretical suppositions that cannot themselves be deemed wholly objective or rooted in a prior, fixed nature. The familiar debate, for instance, about whether the economy should or should not be subject to political direction depends crucially upon prior understandings of both what politics is and what an economy is.

If some people regard the natural world as either autonomous and un-affected by humanity—or as shaped strictly by autonomous forces of technological development or the economy—this is possible only because they first define or interpret these realms as nonpolitical or autonomous of politics. This definitional process is one that itself must be understood as political in the inclusive sense toward which both Hobbes and Aristotle point us. Some constructed definition of the character and scope of poli-tics is inevitable. We saw this inevitability in considering Aristotle's views, where it became necessary to distinguish between the *polis* as the whole of human society and the *polis* as a specific site of political decision-making. Similarly for Hobbes, while sovereignty is said to incorporate all artifice, the sovereign is nonetheless understood to leave large areas unregulated and hence untouched by its power. We are misled if we come to believe that Hobbes and Aristotle can advance such conceptions on any distinc-tively natural basis; it is here that the dualist interpreters are on the firmest ground. In order to clarify the argument advanced to this point, it will be valuable to further distinguish the differing ways in which "politics" is understood here.

Two Senses of Politics

[Although] everything can be the object of political transaction at some point, not everything can be political at the same time.
—Claus Offe[4]

Offe encourages us to distinguish between two senses of politics that have produced a creative tension to this point in the chapter. The first and more general of these is politics viewed as akin to the order of human society itself. It is in this sense that Aristotle, for example, has spoken of the *polis* as natural. "Everything," here, "can be the object of a political transaction." We might say that everything here is *potentially* political. It may seem tempting, in fact, to offer some alternative terminology for this sense of politics (perhaps "culture" or "society") since it is broader and more inclusive than the meanings that we typically attribute to the term when we speak of political processes or debates. This temptation is worth resisting. In the end, recognizing this undifferentiated, naturally constituted entity as a polity is quite valuable. It reinforces our awareness that all human relations and activities, including production, technology, agriculture, familial and gender relations, can be the subject of political choices.[5] Conversely, it should make us aware that the structure of these other realms rules our lives in a way comparable to those more commonly called political.[6] It also allows us to better appreciate Andrew Dobson's contention that environmentalist political thought should promote a recognition that "the natural world . . . is a site of political activity."[7]

As I have shown previously, both Hobbes and Aristotle at times write about politics in this broad and undifferentiated sense. Indeed, presenting politics in this way (as the "master science" for Aristotle, or as undivided and absolute sovereignty for Hobbes) is one of the most familiar, important, and distinctive elements of their political theories. This has, I have noted, led some critics to accuse each writer of harboring totalitarian visions or implications. Neither philosopher, however, is guilty of this accusation. This is because both also offer a second, more constrained vision of a politics that is differentiated from the totality of human activity. Both readily acknowledge that, in Offe's words, "not everything can be political at the same time," and hence that political authority is not going to be felt, and certainly not felt with equal force, in all aspects of life at the same time.

For Hobbes, sovereignty is said to be limited to providing the "hedges" within which humans can act.[8] Within these hedges, however, Hobbes's sovereign ensures the space necessary to foster the pursuit of "commodious living."[9] Hobbes always regards the sovereign as the interpreter of the requirements for this pursuit, but these requirements nonetheless are presented as both natural constraints and a source of guidance for the sovereign within Hobbes's theory. For Aristotle, politics is distinguished from the household realm. While this latter realm is necessarily subject to political decisions, Aristotle also presents it as a realm apart from the public one.

These presentations, it would seem, often blind the theorists (or at least readers of their theories) to the non-naturalness of relationships within the realm of commodious living for Hobbes, or the household for Aristotle. Nonetheless, it ought not be said that they fail in the contrasting respect of not allowing any room for maneuver outside of the ever-present discipline of a totalitarian state. It is because of this second usage of the concept of politics, applied to a specific set of institutions and processes within a broader polity, that the dangers of complete control are seemingly averted. It is with this second usage that we come closer to a contemporary sense of politics as characterized by the particular institutions of government and related institutions and processes.

Each of the two dominant interpretations of Western thought—dualism and derivation—appeals to a different sense of politics. Politics in the inclusive sense is constituted naturally. It is improper to claim that politics in this sense is "derived" from nature, because the delineation of the particular character and scope of a political process is precisely what this undifferentiated sense of politics lacks. Nonetheless, the intuition that politics must be natural is supported by much of the evidence typically used to bolster a derivative interpretation. As we have seen, derivative interpreters wish to say more than this. They wish to affirm a *particular* vision of political order as a distinctively natural one—one distinguished by having been derived from nature itself. For this position, the case is not convincing.

To see why the case for a distinctively natural politics is so weak, I have indicated the differences in the sense of politics upon which it relies. In considering this particular sense of politics, dualist interpreters have advanced the stronger case. Politics, in this sense, is a human creation. It is subject to definition and construction by both theorists and political actors. By being aware of this quality, we can reject the notion that politics in this sense can be justified by an appeal to nature. A consequence is to blur the boundary between political theorists and political actors, prompting us to see that the work of the former as well as the latter involves the construction of a particular vision of the character and scope of politics. This vision separates politics from the undifferentiated whole and defines its relationship to other realms.

Much writing by political theorists can best be understood as political acts. Among the questions that such acts can address are those that focus on the *scope* of the activity. Is it understood as inclusive of economic relations? technological development? gender relations? nonhuman interests or needs? Answers to these questions are not necessarily synonymous with one's moral attitudes regarding such concerns. Nor, as I have argued, can they be defended on the basis of claims about nature itself. They are best understood as political decisions in the broadest sense, although they are only sporadically the subject of attention by institutional political actors. For political theorists, however, the delineation of the boundaries between these realms tends to be much more central. With careful attention, we can gain an appreciation for the inescapably contingent quality of this process.

In a similar manner, the definition of the *character* of politics is a contingent one. Is politics (or can it be) deliberative or aggregative? participatory or oligarchic? more comprehensive or more limited in its ability to anticipate and respond to complexity? Disagreements here also revolve around whether politics is understood as seeking to direct citizens toward a particular, substantive, conception of right and virtue, or whether it is conceived as limited and prudential in its orientation. Of course, normative arguments about such views abound and I have reviewed a number in this book. By recognizing that there is no natural basis for resolving these debates, we can recover the interpretive dimension of politics.

The images offered by Aristotle and Hobbes, of the natural order or disorder that precedes politics, are not the reflection of their conception of nature itself. While natural principles play a key role in legitimating them, the particulars are shaped by the theorists' conceptions of the boundaries and character of politics itself. Thus, the "natural" condition described by these authors is a political construction, not a consequence of their respective conceptions of natural teleology or mechanism. While this is vitally important, we must not allow it to eclipse the fact that these theories also allow us to recognize the effect of the natural world in shaping our politics and some of the ways in which human activity affects the natural world. We must challenge the fusion of a relationship that is true on a physical, material, constitutive level with an ideational relationship that attempts—but must ultimately fail—to provide natural, directive principles for politics. Nature in a physical, material sense is constitutive of human endeavors, including politics, precisely because humanity cannot be truly divorced from the world in which it exists. Conversely, and for precisely the same reason, politics—understood as inclusive of all human endeavors—shapes and alters the world around us.

A Political Nature

In contrast to the dualist view, a dialectical relationship between conceptions of nature and politics avoids beginning with an untenable premise that ignores our place in the world. In contrast to the derivative view, a dialectical relationship rejects the premise that nature can authorize the particular organization of politics or other human institutions. We must recognize the ways that conceptions of nature and politics interact even when theorists fail to make this explicit.

The product of the dialectic that I am describing is a conceptual category that is typically characterized as "natural" and yet delineated politically. As such, it can be appropriately termed a *political nature*. By mediating between conceptions of nature and politics, this category partakes of some qualities from each. When we ignore or suppress the political qualities of this mediating concept, it appears to take on an authoritative and lawlike relation to the question of the proper ordering of human communities. It is in this manner, we have seen, that many discussions of Aristotle and Hobbes are framed. It is also in this manner that appeals to an ecological worldview are quite often understood. Conversely, when we dismiss the natural qualities of this conception, we naïvely allow ourselves to treat politics as unencumbered by our inescapable connections to the world. Thus we reinforce the dualist reading of Western thought.

Because the mediating role of this political nature is unavoidable, there are good reasons to bring it to the surface. The most important reason is that in practice, when we ignore or submerge this dialectic, we relinquish power to others who have a stake in constructing our political conceptions.

If nature is understood as presenting us with answers to political questions, then the rise of environmental concern and environmental politics is likely to be greeted as a basis for the adoption of such answers. To the extent that this understanding of nature remains influential, it is likely to offer justification and cover for changes that allow a powerful minority—of individuals, nation-states, or both—to adapt effectively to new environmental conditions while doing little to address the interests or concerns of many others. The hopeful visions of greater participation, democracy, and equality that many environmentalists advocate will get little serious attention in this case. The argument of this book, of course, does nothing to change these power relationships. However, a greater awareness of a dialectical relationship could help to undermine the use of nature as a source of legitimacy by elites, while strengthening the argument in favor of confronting the questions raised by our relationship with the natural world.

My appeal to "dialectic" may invite ambiguities and misconceptions since the term carries significant baggage and diverse connotations. One connotation from which I wish to distance my argument identifies dialectic as a dynamic of social change or historical progression. Hegel and Marx, of course, confidently interpreted the past and charted out the future by defining the dialectical relationship of *geist* or materialism, respectively.[10] Used in this manner, dialectic can obfuscate rather than clarify the basis for a commitment to a particular vision of social change.

As I use the term, dialectic is intended to be more straightforward and less powerful than this. The recognition of a dialectical relationship between conceptions of nature and politics makes clear their interactive participation in the delineation of a third conception that I have termed a political nature. However, it does not offer definitive answers to normative questions, nor does it allow us to forecast the future. Instead, our awareness of it allows us to identify a task—a debate—that ought to be central for those of us concerned about the character of the world of which we are a part and the human societies of which we are members. I wish to resist the determinism to be found in Marxian and Hegelian versions of the dialectic, as well as the determinism I have identified in the dualist and derivative approaches.

The existence of a dialectical relationship between conceptions of nature and politics prompts us to see that determinants of the character and scope of the political sphere in relation to other spheres of human action are left unresolved by a commitment to any conception of nature. For the committed theorist, this lack of resolution is messy and can also be disappointing because it forces us to recognize the limits of theory. By contrast, both the derivative and the dualist interpretation appear to offer the possibility of a wholly theoretical and ultimately self-contained resolution. Politics, divided from nature, has an eternal and static quality to it. While in practice it might often be contaminated by its interaction with nature, in theory we could imagine its pure and unchanging, transcendent form. Politics, derived from nature, also has a fixed quality. Again, this position does not mean that we imagine such politics as always or even often existent. But in theory we could identify a political ideal that adheres to natural principles. As long as our conception of nature is believed to be correct, this ideal will also remain unchanged.

No such theoretical resolution is possible for a dialectical relationship. While a close attention to theory has helped to reveal this relationship, only less eternal, less transcendent forms of inquiry can authorize (even temporarily) the character and scope of politics in relation to other spheres of human action. Political judgment becomes a necessary virtue in this situation. As Ronald Beiner has argued in a parallel context,

in so far as they claim to supply universal rules or principles for determining all judgment, such theories situate politics within the sphere of determinant judgment, and fail to capture the dimension of reflective judgment that characterizes the world of human affairs as something not calculable but essentially dramatic.[11]

Certainly, the exercise of such "reflective judgment" need not be antitheoretical, nor even atheoretical. However, the theorizing with which it is compatible is best conceived as social or political criticism[12] grounded, not in the supposed absolutes of nature, but in our experience of the particularities of place.

Focusing on this dialectic suggests the importance of a range of topics that are far from alien to the history of political thought yet are rarely central to contemporary debates. These include the planning of public and private spaces, patterns of land use, the role of human and physical geographies, and the relationship and impact of technologies upon both human and nonhuman communities. Aristotle recognized such questions as central in his delineation of the *polis* in a manner uncommon among recent political theorists. At the same time, however, we have seen that the attention of many readers is diverted because Aristotle limits human interaction with the world to the realm of household management. Hobbes, too, recognizes the ultimate dependence of all these endeavors upon politics. To the extent that they fall outside the sovereign's function of ensuring peace, however, they are described as part of the pursuit of commodious living; a pursuit defined as central to, although stifled in, the "natural condition" and hence naturalized itself.[13] The consequence is to remove it from active political consideration.

Consider, for example, contemporary discussions of "sustainable development." This term has gained significant prominence in the past decade and has become one of the most popular frameworks for discussion of environmental issues in both the so-called third world and the first.[14] Notably, it has been adopted not only by environmentalists, but also by government and business elites and international lending institutions.

But what does "sustainable development" mean? What is it that should be *sustained?* an ecosystem or forest? a species within an ecosystem? a way of life for human communities? development itself? and what sort of practices and actions can ensure this sort of sustainability? government ownership of the land? local, private enterprise? regulation of corporate behavior to protect endangered species? legal recognition of indigenous land tenure arrangements? management of resources by nonprofit environmental groups? These questions are further multiplied once we ask about the meaning of *development.* "Sustainable development" clearly means different things to different groups and individuals.[15] Unsurprisingly, the interpretation of the term adopted by the World Bank (for example) has been more influential in shaping policies than the interpretation of many radical environmentalists.

Why might this be so? The perspective advanced by many writers considered critically in this book suggests that (to use this simplified example) the views of the World Bank are influential simply because most people adhere to a worldview that prizes economic production above all else and fails to recognize our ecological interdependence. Others have argued that terms such as "sustainable development" are coopted by powerful, self-serving elites seeking to "greenwash" their activities while avoiding more fundamental changes.[16] Neither explanation is without some merit. Neither, however, is complete. A recognition of the power of institutions such as the World Bank is necessary to explain the influence of their views, but characterizing this as a "cooptation" of a concept such as sustainable development fails to recognize the *inevitability* of interpreting the concept in some manner. Many do use this concept cynically for their own ends. Even in the absence of such motivations, however, there is no fixed meaning, no fixed understanding of the proper relationship of public and private; market and planning; governmental and nongovernmental activities that will lead to "sustainable development."

Even if it made sense to speak of a community composed entirely of adherents to an "ecological worldview," they would continue to be confronted by differences of interpretation and judgment on these matters. The advantage held by such a hypothetical community over actual ones is that they would seem forced to recognize the role of these judgments—and hence to delineate, defend, and debate them—since there would be no basis for dismissing one's opponents as inadequately "ecological," or as adherents of an incorrect worldview. While the diverse problems that have led to a debate over sustainable development are posed to us as a result of our being part of nature, the manner in which we ought to address them is not to be found there.

The way we shape the boundaries and realms of human life are political decisions that affect our understanding of what the natural world is and how we as humans interact with it. Moreover, these boundaries affect the way that we understand all those realms of human life within which our interaction with this natural world occurs. Thus these boundaries themselves help to shape the natural world. This is one side of the dialectic. The other side is the fact that this natural world shapes who *we* are—precisely the "we" that construct the boundaries in the first place. The emphasis here is very much on material relationships rather than ideal conceptions.

Nature, Experience, and Place

In a penetrating essay that invited the development of a pragmatist perspective within environmentalist thought, Anthony Weston argued that scholars of environmental ethics have been unjustifiably preoccupied with finding a source of "intrinsic value" in nature.[17] "The problem," he contends,

is not to devise still more imaginative or exotic justifications for environmental values. We do not need to ground these values . . . but rather to situate them in their supporting contexts and to adjudicate their conflicts with others—a subtle enough difference at first glance, perhaps, but in fact a radical shift in philosophical perspective.[18]

Weston maintains that providing philosophical justification does little or nothing to protect or bolster environmentalist values. He offers two reasons for this. First, claims for intrinsic value in nature offer little guidance in the myriad situations where conflicts between values must be resolved. The attribution of this type of value to all beings, species, and ecosystems virtually guarantees that conflict will not be resolved on the basis of claims of intrinsic value alone. Second, environmentalist values are rooted in experiences that often can be only inadequately conveyed through language and especially philosophy.[19]

By emphasizing this experiential and interdependent character of our values, Weston's pragmatism also reveals the actually existing basis for fostering environmental concern. As he explains,

Nearly everyone recognizes some value in nature. . . . Even motorboaters like to see woods. Wilderness values may just seem to them less significant than other values at stake in the particular situation. . . . *the real issue shifts to the question of alternatives.* . . .[20] [emphasis added]

Mark Sagoff addresses this question of alternatives through his defense of incorporating "public values" concerning aesthetics, health and safety, natural heritage, and respect or reverence for nature into the political process.[21] He argues for the necessity of political deliberation in order to adequately incorporate these values. This deliberative process is contrasted with policy-making based narrowly on cost–benefit calculations in order to bolster the legitimacy of noneconomic claims made by environmental advocates in political debates. Key to his argument is the conviction that the existing public values, at least in the United States, are strong enough to support a firm commitment to an environmental politics.[22]

The problem identified here is neither a Western worldview nor existing public values, but a restrictive conception of the appropriate justifications for politics and public policy. Sagoff's argument illustrates the role played by alternative political conceptions in delineating environmental problems and solutions. By doing so, he challenges the premise that competing yet unified worldviews rest behind the political positions adopted in environmental discourse.

What, then, can we understand as the sources of the category of experience that plays such an important role here? Timothy Kaufman-Osborn explains,

It is crucial that [experience] be understood in relational terms, that is, as an articulation of the way nature's dynamic events come to be incorporated within the equally mobile realm of human being.[23]

Experience, as he argues, partakes of both human artifact and nature, where the latter is understood as outside the realm of human construction. It is the product of a dialectical relationship between the two. This insight into the character of experience can be carried over into an understanding of another category as well—the concept of *place*. A place emerges, Sagoff argues, "when it is cultivated, when it constrains human activity and is constrained by it, when it functions as a center of felt value because human needs, cultural and social as well as biological, are satisfied in it."[24]

Place is key because it *is* the "environment" in the relevant sense so often contested politically. Place and experience are constituted through, and serve to mediate, the nature–politics relationship. They are the *political nature* that is the basis for, but not the authoritative principle of, an expansive environmental politics. This environmental politics consists of our struggle over the creation, use, preservation, alteration, and degradation of place. This struggle is defined by our relationships to these places and our experiences in them, in all their complexity and diversity.

The point in stressing the importance of place is emphatically not to argue that familiar preservationist concerns ought to be discarded from the environmentalist agenda. The point is to recognize the common components in these efforts and in struggles over other places less likely to be described as "natural." As some scholars have recently indicated, even the seemingly fixed and sacrosanct concept of "wilderness" is a *place*. While it holds much power as a preservationist ideal, when presented as *nature* it can be used to obscure or ignore long-standing relationships between people and the land.[25]

A politics of place would not be restricted to the local. Our relationship with places is often affected by structures, institutions, and technologies that are national or global in scope, so our political thinking and engagement must operate on these scales as well. Some might argue that a politics of place is increasingly obviated by the number of people for whom the growth of computer, communication, and transportation technologies have made a cosmopolitan and even "virtual" existence commonplace. Yet the presumption that place is relevant only to a bygone era, in which most people's geographic horizons were limited, is misguided.

Cosmopolitanism and the emergence of a virtual culture alter the landscape and alter our perception of the physical places in which we live. New technologies can be central to the definition of place and in many cases to the perceived homogenization of places.[26] Still, there is no substitute (re*place*ment) for the centrality of place in our lives, as long as we remain part of the natural world. We can ask how these technologies visually affect the places in which we live, how they affect our patterns of daily life, how they affect our health and well-being and that of our children. Most often, such questions are regarded as marginal at the level of theoretical inquiry into politics (as well as marginal to the politics of nation-states and international relations). Yet it is these issues that are key to who we are and what we value.

In the end, then, what might an environmental politics based on experience and place look like? It cannot convincingly offer a noncontingent argument for a particular political process or content—politics in the exclusive sense—precisely because it recognizes the potential politicalness of all forms of societal interpretation and decision-making. Committed environmentalist thinkers have argued for and against democracy, for and against authoritarianism, for and against anarchism, for and against free markets, for and against technology, for and against civil society, for and against religion. In the process of developing interpretations and making judgments about these matters, we are in effect selecting particular ways of conceptualizing our relationship to the natural world. Only if we recognize this can we appreciate the inescapability of the inclusive sense of politics. In this sense, we must politicize the project of political theory itself.

Many have argued that in a world without certainty, in which the sort of fixed, foundational claims of legitimate authority tendered by many environmentalist thinkers have been challenged, democracy is the only source of authority to retain legitimacy.[27] There is real insight in such arguments. Certainly my argument in favor of politicizing political theory can be appropriately read as an effort to *democratize* this practice, where by this I mean to open up such constructions to broader and deeper discussion and deliberation. Nonetheless, by distinguishing between an inclusive and exclusive sense of politics, we can also recognize that democratizing the former does not necessarily lead to the establishment of particular democratic political processes in the latter. The very meaning of democracy in this latter sense will be dependent upon the definitions and interpretations advanced prior to this process.

The environmental politics that I am pointing toward would not be limited to the advocacy of a particular form of government, or to involvement in any single arena of activity. It would certainly involve governmental and legal policy-making, but it would also focus on opportunities to raise environmental concerns and to promote changes in corporate behavior and market activities and private lifestyle and consumption patterns. By doing so, we would not be ignoring the significant and undeniable importance of governmental politics and policy, or retreating into the ambiguous realm of "consciousness raising." This environmental politics would be acting on the recognition that the boundaries of all spheres of human activity are constructions subject to reinterpretation and change. All would be *potentially* political—whether or not they ought to be subject to explicit government directives. The notion, for example, that environmentalist values are alien to economic activity while exploitative values are at home here might be challenged by consumer boycotts as well as by congressional lobbying. Which strategy might be more effective is contingent on the particular case and cannot be determined in a theoretical argument.

In recent times, environmental concerns have been widely shared by citizens across the globe.[28] Yet environmentalist writers have often ignored the existence of these shared concerns, or else belittled them as shallow and insignificant. They have relied far too heavily upon an unconvincing view of the environmental movement as *conceptually* marginalized.[29] Understandably frustrated by the frequent inability of our practices to adequately address these concerns, many persons have focused on promoting a new conception of nature as the answer to environmental problems. The fact that widespread concern can and does exist concurrently with inadequate changes in practice should tell us that this focus is misplaced.

Whether we consider the deforestation of the Amazon or the air pollution in Los Angeles, we must look at the many ways that implicit judgments about what is public and what is private, what is political and what is economic, shape our ability to even imagine the possibilities for addressing these problems. Within the realm of governmental activity, the presence of environmental values has at times been significant. Yet, faced by powerful agents of economic and technological change and dislocation who are not guided by these values, legislation and regulation are often half-hearted and ineffective.

Why are these agents—whether corporate landowners, government road builders, or military institutions—so often judged to be outside the realm in which environmental concerns can be addressed? Why is environmental action so often limited to a narrow area of government regulation and private voluntary actions? These are questions that must be central to environmental analyses. The answers clearly direct us toward evaluating the relative power of economic, social, and political actors.

The dialectic between nature and politics suggests that the struggle to configure a variety of human institutions and realms in a way that is open to environmentalist values and concerns will be a continuing one. The very process of doing so will alter the ideas, attitudes, and behaviors of those involved on all sides of this struggle in ways that will influence future engagements. There is no end to this dialectic, as there is no end to our participation in the natural world itself. The illusion that such an end is possible may be dangerous and certainly distracts us from the task at hand.

7

New Possibilities for Environmental Politics

We seek not some philosophical "grounding" but the actual ground: the Earth.
—Anthony Weston[1]

While I hope that readers have found my theoretical argument persuasive, many people sympathetic to environmentalist concerns may remain uneasy, worried that this analysis makes one an untrustworthy proponent of these concerns. In this chapter, I seek to address this worry. My focus is on what might be gained as a result of the supposed concession that nature cannot serve as an independent source of authority for political action. I do so primarily by considering three cases that frequently pose uncomfortable challenges to received notions of environmentalism: the "environmental justice" movement in the United States, ecological resistance movements in third world countries, and the "land rights" movement in the western United States. Despite crucial differences among these cases, all three build upon the centrality of experience and place for environmental politics.

Before developing an analysis of these movements, I wish to relate my argument to the more familiar environmentalist commitment to the protection of wilderness and other "natural" places. This commitment has been central to the most prominent environmentalist organizations in the West *and* to those critics who identify with deep ecology and other ecocentric views. This is especially significant because many concerns about a critique of nature's authority rest on the fear that it will undermine protection efforts of this sort.[2]

Advocates of wilderness preservation and deep ecology often emphasize the importance of experience in a particular kind of place. Preservationist efforts are often defended with the argument that experience in "natural" places can temper the negative effects of the constructed, "civilized" world. John Muir made this explicit in an oft-quoted passage from over a century ago:

Thousands of tired, nerve-shaken, over-civilized people are beginning to find out . . . that mountain parks and reservations are . . . fountains of life. Awakening from the stupefying effects of the vice of over-industry and the deadly apathy of luxury, they are trying as best they can to mix and enrich their own little ongoings with those of Nature, and to get rid of rust and disease.[3]

While Muir's emphasis here is on the role of these parks as places beneficial for people, his goal can also be viewed as the creation of places within which preservationist commitments are likely to be developed.

Similarly, advocates of deep ecology frequently share the experience of identification with the nonhuman world in a manner that fuels their ecocentric passion. Thus Warwick Fox has argued that

the fact—the utterly astonishing fact—that things *are* impresses itself upon some people in such a profound way . . . that "the environment" or "the world at large" is experienced not as a mere backdrop . . . but rather as just as much an expression of the manifesting of Being (i.e., of experience per se) as we ourselves are. . . . If we draw upon this experience we can then gain some insight into why it is that people who experience the world in this way . . . find themselves tending to experience a deep but impartial sense of identification with *all* existents.[4]

One way to obtain more advocates for such a position might be to develop programs that foster and enable these experiences among a wider segment of the population. However, as Fox himself seems to acknowledge, we cannot prescribe the experiences that people actually have in the places where they live, work, and play.

The recognition that we must begin from our experiences distances us from philosophical debates between anthropocentrism and ecocentrism—and other similar dichotomies—by a pragmatic acknowledgment that different individuals and communities will draw upon their diverse experiences in an eclectic manner. In the absence of relevant experience, no analytical argument for ecocentrism will make the identification that Fox describes convincing. In the presence of the experience he describes, the question turns to the relationship between it and the inevitable variety of other experiences that we have. For example, an emphasis on experience illuminates but does not resolve the potential conflicts between the perceptions of place that can emerge through our work and those developed through play.[5] It does, however, open environmentalism out beyond a concern with wilderness preservation and the like and thus allows the concerns of the movements described later to fit more comfortably within its newly expanded boundaries.

Conceptions of place and experience affect our understandings of nature and politics, whether the issue is the protection of "wilderness" or the reduction of toxic hazards in the workplace. There is strategic value for environmentalists in advancing this argument because it has the potential to broaden the movement and its constituencies. Yet my argument is not prompted by considerations of strategy in this limited sense. Of greater importance is that an environmentalism understood in this manner can avoid the dangers of reifying nature as the standard for political action while remaining true to the deeply felt concerns and convictions that motivate the work of those who seek to protect, preserve, or improve the quality of the environment in which we all live.

In the following three cases, I illustrate ways in which participants' experiences and sense of place mediate between their relationship with the world, on the one hand, and their political commitments on the other. While I limit my discussion here to the examples at hand, I explore their implications more fully in the final section of this final chapter.

The U.S. "Environmental Justice" Movement

The "environmental justice" or "grassroots antitoxics" movement in the United States has in recent years highlighted racial and class disparities in the distribution of environmental hazards such as toxic waste dumps, polluting factories, and poor air quality.[6] Its membership has been far more diverse than that of other groups in the contemporary American environmental movement. In both its tactics and its rhetoric of social justice, this movement has also been among the most radical and uncompromising elements of the environmental movement. Strikingly, its membership and even its leadership has been wary of identifying themselves as environmentalists and is sometimes disdainful of appeals to a nature that transcends their concerns with the health and safety of their (human) communities. As prominent leader and activist Lois Gibbs puts it,

Calling our movement an environmental movement would inhibit our organizing and undercut our claim that we are about protecting people.[7]

Gibbs's explicit human centeredness has been challenged by others who advance a more inclusive understanding of the relationship between their struggles and those of other environmentalists. For example, the first of seventeen "Principles of Environmental Justice" adopted by the First National People of Color Environmental Leadership Summit, "affirms the sacredness of Mother Earth, ecological unity and the interdependence of all species, and the right to be free from ecological destruction."[8] Nonetheless, this movement cannot be understood as one reliant upon an appeal to nature. Thus if we use the now-familiar dichotomies in environmentalist theoretical work, the only category within which this movement can be understood is one labeled "reformist" or "mainstream" environmentalism. Yet this label is clearly inappropriate for a movement whose priorities, tactics, and membership all diverge significantly from that of the large, established, national environmental organizations in the United States. Given these limited choices, Robert Gottlieb's claim that this movement represents "a new way of defining what it meant to be an environmentalist" makes sense.[9]

The basis for the concerns of this movement's members can be found in the places where they live and work. It is here that exposure to environmental hazards leads to activism intended to alter these conditions:

They see their communities used as a dumping grounds for toxic wastes because they lack the power to prevent it; they find only the most hazardous jobs available, if they can find jobs at all; and they discover that the circumstances of their work are set by managers and engineers who do not have to suffer the hazards of the job floor.[10]

By definition, experiences rooted in the particularities of place cannot be divorced from nature and cannot be adequately described within a dualist framework. Places are a complex interweaving of the biophysical and the humanly constructed. Yet the qualities of the communities that we call home and the factories, fields, or offices in which we work are also quite clearly not protected by any appeal to natural authority.

Given the movement's priorities, the racial and class diversity of the environmental justice movement in the United States is easily comprehensible. The gender composition requires a bit more attention. In seven of the nine case studies discussed in one pathbreaking book on environmental justice organizations, it was women who were most prominent in organizational leadership roles. "[T]he typical grassroots leader was a woman. . . . Women activists were quick to express their concern about the threat to their family, home, and community," Robert Bullard observes.[11] Also significant is that the women taking on these leadership roles were ones who were often new to political action and to public roles more generally; it is, some of these activists have asserted, a "movement of housewives."[12]

What can we learn from these observations? Clearly, the motivations that drive many into these leadership positions are distinctive. "We become fighters," explains activist Cora Tucker, "when something threatens our home."[13] Rather than being prompted primarily by a scientific understanding of nature or environmental threats, rather than being dependent upon something as abstract as the transformation of one's worldview, the leaders of this movement are galvanized by concern for the seemingly most familiar elements of daily life: children's health and the health and vitality of their neighborhoods. The issues are seen in terms of personal, familial, and communal experience and as issues involving the importance and protection of place.[14] As Gottlieb has noted, "The importance of place has become a dominant and powerful metaphor within the antitoxics movement."[15]

Yet place is more than a metaphor here. It is the grounds of experience, the location of family and friends, the basis for one's livelihood. An appeal to place does have rhetorical appeal. It also has significant conceptual power, as I elaborate later. This power and appeal of place as an *idea*—a conceptual category—is the direct consequence of the power and appeal of place as a *lived reality,* in this case for the members of groups such as the "Mothers of East Los Angeles," the "West Dallas Coalition for Environmental and Economic Justice," and the "Gulf Coast Tenants Organization."

The environmental justice movement in the United States is also distinguished, as its name suggests, by its emphasis on questions of distributive justice as key to any resolution of its members' concerns. Bullard makes this clear by contrasting this movement with other manifestations of environmentalism:

Environmental philosophy and decisionmaking has often failed to address the "justice" question of who gets help and who does not; who can afford help and who cannot; why some contaminated communities get studied while others get left off the research agenda; why industry poisons some communities and not others; why some contaminated communities get cleaned up while others do not; and why some communities are protected and others are not protected.[16]

The questions as well as the answers raised here are inescapably political. To the extent that environmental philosophy has ignored these questions, it is largely because it has ignored the influence of political judgments on its conception of what counts as an environmental problem and what counts as an adequate or appropriate solution to these problems. The environmental justice movement is far more attentive to these influences.

Still, attentiveness to questions about distributive justice and to the political and power relationships that underlie issues of distribution does not generate any clear answers to these questions. How does the environmental justice movement propose that the inequities toward which they direct our attention be resolved? Is stronger national regulation the answer? decentralization of political authority? citizen pressure on corporate decision-making? Unequivocal answers do not easily present themselves here.

Critics who have sought to discredit the movement suggest that they are nothing more than victims or perpetrators of a "NIMBY (not in my backyard) syndrome," arguing against the use of their own "backyard" for undesirable activities, but lacking any broader vision for how the environmental hazards might be addressed. Such a dismissive claim is unconvincingly critical; even where such a characterization of their motivation is appropriate, it is obnoxious to impugn the motives of people acting upon a perceived threat to their family and community simply on the grounds that their concern is not abstract or "disinterested."[17] It is also inaccurate to presume here that "self-interest"—or more appropriately, localized communal interest—is in conflict with a broader public good. Yet while the derogatory connotation of the NIMBY accusation does not stick, it remains true that the environmental justice movement has not achieved the very difficult task of articulating a complete political vision. Often, as Gottlieb concludes, these groups are:

operating primarily in the context of survival politics . . . without a developed theoretical framework, with limited strategies, and without a longer-term vision.[18]

Of course, we can step back from the movement's activities and attempt to draw some conclusions. Mark Dowie, for instance, convincingly concludes that a distinguishing characteristic is a "strong belief in the right of citizens to participate in environmental decision making," a belief that he suggests is advanced primarily through changes in governmental policy, such as community right-to-know laws, and provisions for citizen lawsuits in the land, air, and water quality legislation adopted on both a federal and state level.[19] Lois Gibbs offers a different emphasis:

Our aim is to change the discussion within the boardrooms of major corporations. . . . That's where we will win ultimately, not in government agencies or Congress. Our strategy is basically like plugging up the toilet—by stopping them from opening up new landfills, incinerators, deep-well injection systems and hazardous waste sites.[20]

The resolution of any debate over visions of how social change should be achieved, whether by direct pressure on corporate behavior or strategies for legal or legislative change, would inevitably be contingent on specific issues. This is as it should be.

Third World Ecological Resistance Movements

While the environmental justice movement poses an important challenge to the conventional wisdom about the postmaterialist character of environmentalism in the United States, the existence of ecological resistance movements throughout the third world offers an equally important challenge on a transnational level. These movements are typically composed of peasants, workers, indigenous peoples, and other disenfranchised groups that struggle to protect the ecological resources of a region. To assess the character of this challenge, we must first appreciate the basis for these movements themselves.

Perhaps the best way to gain an understanding of third world ecological resistance movements is to begin by seeing what they are not. Bron Raymond Taylor has carefully studied and written about the values, narratives, and activities of Earth First! activists in the United States. These activists identify with deep ecology and have become well known for their participation in forms of civil disobedience as a means of protecting old-growth forests and wilderness areas. Earth First!ers frequently identify movements in the third world as kindred movements and assume that their own ideals and bases of resistance are consistent with those of groups abroad. Taylor began studying these third world movements as a way of assessing the accuracy of these assumptions. He seems to express surprise at discovering "significant discontinuities" between the two sorts of movements.[21] Summarizing the findings in the book he edited, he states that

popular ecological resistance [in the third world] often originates in a desperate quest for survival as industrial processes threaten habitual modes of existence and as people recognize that their well-being is threatened by environmental degradation. . . . threats to human livelihood and health provide the most important reasons for the global emergence and proliferation of popular ecological resistance.

This strikes a rather different chord than that expressed by Earth First!ers. As Taylor notes, this motivation:

might surprise those radical environmentalists who believe that a *transformation of consciousness,* from anthropocentrism to biocentrism, is a prerequisite of ecological resistance.[22]

What has been termed the "environmentalism of the poor" in the third world turns out to have more in common with the environmental justice movement in the United States than it does with activists such as Earth First!. At the core is a conflict over the distribution of ecological resources and a struggle to preserve those necessary to ensure their livelihood and community.[23] Ramachandra Guha characterizes the movements as a "defense of the locality and the local community" by those who live there.[24] Among the best-known examples of this sort of movement is the struggle by peoples in the Brazilian Amazon to protect the rainforest. The goal of this struggle has not been the preservation of a wilderness area or the creation of a national park. Instead, it has been the creation of "extractive reserves." Chico Mendes writes:

What do we mean by an extractive reserve? We mean the land is under public ownership but the rubber tappers and other workers that live on that land should have the right to live and work there. I say "other workers" because there are . . . people who earn a living solely from harvesting nuts, while there are others who harvest babaçu and jute.[25]

The extractive reserve is intended to protect the Amazon from the devastating effects of corporations, cattle farmers, road builders, and landless peasants who are clear-cutting wide swaths of it. The rubber tappers and other forest inhabitants promoting the concept have lives and livelihoods that are intimately linked to the continued existence of the rainforest in all its diversity. Yet their ecological interests are inextricable from their economic ones; their defense of the rainforest is also a defense of their work and lives in the rainforest:

We, the Indigenous Peoples, have been an integral part of the Amazonian Biosphere for millennia. We use and care for the resources of that biosphere with respect, because it is our home, and because we know that our survival and that of our future generations depend on it.[26]

Some may raise legitimate questions about whether the characterization of indigenous practices is romanticized here. What is clear, however, is that this is once again a defense of place—and a community's experience in that place—against those interests perceived as threatening it.

Opposition to both the state and the influence of multinational corporations typically characterizes the environmental struggles in the third world.[27] Both entities often play a key role in the enclosure of what had been a "commons"—land perhaps nominally owned by the state, but long utilized and managed by local people as the basis for their livelihoods—into property that is transformed in its character (into an oil field, or cattle ranch, or monocultural farm, for example) and also exploited by different beneficiaries (corporations, national elites, or other "outsiders"). This struggle over places used as a commons is a distinguishing characteristic of ecological resistance in the third world.[28]

The "Land Rights" Movement in the American West

While the movements reviewed in the preceding sections suggest the potential for an environmentalism rooted in experience and place, the contemporary "wise use" movement and other manifestations of the broader "land rights" movement in the western states of the United States seem to offer a very different perspective. The first two movements can be best understood as offering an internal criticism: challenging the boundaries of environmental politics and redefining environmentalism. By contrast, the land rights movement in the American West defines itself by its opposition to environmentalism as they understand it.[29] Its proponents attack a movement that they view as not just outside, but largely antithetical to their own interests.

A coalition of "loggers, ranchers, miners, farmers, fishermen, oil and gas interests, real estate developers and off-road vehicle enthusiasts," the movement has often been supported and funded by corporate interests.

Because of this, many environmentalists have dismissed it as nothing more than a front for powerful natural resource industries.[30] However, it should not be dismissed so readily. As Donald Snow puts it,

Wise use, daffy as it often sounds, is at the core a pleading to maintain, somehow, somewhere, a dignified rural existence. . . . Lots and lots of folks back home, in little hamlets like Deer Lodge and Burley and Salida, don't really want the new economy of environmental amenities. They want the old one, because that's where they find dignity, livelihood, and the completion of what they and their ancestors saw as destiny.[31]

Like both of the other cases considered here, "wise-users" are committed to the places in which they live and work. Environmentalists have often, and rightly, argued that the economic well-being of the western states is no longer primarily dependent upon the natural resource extractive industries that wise-users are so often a part of and that it has not been primarily environmental policy but broader economic changes that led to declining employment in these industries.[32] Yet by itself, this response fails to address the claim being made by this movement, which centers on their concern with "livelihood and equity."[33] It may be that many who support the movement could find secure employment and housing by relocating to sprawling and anonymous suburbs that surround contemporary American cities. However, the emotive basis for the land rights attack upon environmentalism is based less on economics than on the quality of life possible in their communities and the perceived threat posed to this life—not just their income—by environmental policies. As Snow argues, "the central problem with environmentalism is that it lacks a cogent, convincing focus on livelihood, and that has made it vulnerable to wise use attacks."[34] Like the environmental justice movement and third world ecological resistance movements, the wise use focus on livelihood is a reflection of the roots of its proponents in the places from which they emerged.

Consider, for example, some striking parallels between the land rights movement and the third world environmental movements. Both assert that their origins are in a concern for the vitality of rural peoples and communities. Both movements have also been dismissed by others as relics, left behind by the increasingly globalized economy. Both have resisted the imposition of distant government authority over local land and land uses in the name of what some activists in the western United States describe as "community stability."[35] It is interesting that both movements have also, at least selectively, resisted the application of "free-market" economic policies to their local or regional economies:

the wise use movement never got off the ground until environmentalists seized the economic initiative and began attacking federal subsidies, both hidden and overt, that prop up the natural resource economies of the West. . . . When environmentalists finally got around to the federal budget, wise use organizing went ballistic.[36]

Of course, economic self-interest plays a key role in this resistance. Yet this position can be understood more broadly as an attempt to secure the continuity of their communities and their relationship to the places where they live, in the face of the abstract and insensitive forces of both the marketplace and the state.

Contrasts between these movements are also evident. Their political positions are radically different. Unlike third world movements, wise use opposition to the elimination of federal subsidies in favor of the market is a position that unites them with the resource extraction corporations that so often dominate their regions. For many, the community that they seek to protect or "stabilize" is defined in large part by the large-scale industries that drew them or their ancestors to their present location. From this perspective, these industries are less an imposition upon their place than a central feature of the landscape.

A second important contrast is the centrality of an exclusive conception of private property rights to the land rights movement. Appealing to a Jeffersonian vision of agrarian land ownership as the basis for both individual autonomy and civic virtue, this movement has drawn heavily upon a literalist interpretation of the "takings clause" at the end of the Fifth Amendment to the U.S. Constitution to validate their notion of absolute private property rights.[37] Thus, despite their misgivings about so-called free markets, the land rights movement has embraced a political vision based on an absolutist conception of private property, an acceptance of an expansive role for resource extractive corporations, and a correspondingly minimalist role for a distant federal government. The first two of these, in particular, represent a dramatic departure from the position espoused by third world movements; all three, of course, can be seen as conflicting with the dominant tendencies of the American environmental movement.

Expansive Environmentalism

When nature serves as the orienting concept for the definition of environmentalism, the role played by experience and a defense of place is easily misunderstood. The result is that struggles such as those described here are sometimes not understood as manifestations of environmentalism. After all, these campaigns often focus on issues of living and working conditions that seem at best indirectly or only conditionally related to protection of nature.

Andrew Dobson offers one of the most thoughtful cases for differentiating efforts on behalf of environmental justice and sustainable development from the more familiar manifestations of environmentalism:

"traditional" environmentalism just is about wilderness, resource-conservation, and so on, and while there are clearly points at which the human justice and environmentalist programmes will come into contact (precisely those points, indeed, highlighted by the environmental justice movement), it is too much to expect that these programmes will be (or should be) synonymous. . . . while the paths of these two movements will cross at important points, they by no means always travel the same road.[38]

He also argues that we must be careful when speaking of the environmental justice movement as broadening our understanding of environmentalism:

we need also to remember that it is *narrower* than preserving the environment, in that not all the environment will be "preserved" through the medium of environmental justice.[39]

Dobson is surely right to reject the possibility of subsuming the goals of one movement within that of the other. This cannot succeed, for reasons that he outlines. Yet reorienting our understanding of environmental politics in the manner I outline here is not meant to value one movement's goals over those of another. It is intended to recognize that they share a struggle rooted in experiences of place, while also recognizing that the variety of their experience can lead to divisiveness or conflict as readily as consensus.

In the case of the land-rights movement, the likelihood of a ready reconciliation with environmentalist groups seems relatively remote. Yet even here possibilities for dialogue can be (and are being) created. What is needed is for both groups to recognize their common interest in the future of the place that serves as the source of their often disparate experiences. If this condition is met, then environmental politics, with all the contentiousness that this term ought to entail, can be recognized as central. Where and when it is not met, we are left with little more than two parties talking past each other at a distance[40] and a politics that can be little more than a façade for the raw exercise of power.

In the introduction to his major new history and interpretation of American environmental policy, Richard Andrews makes the following argument:

Environmental policies are intricately interwoven with the broader forces and patterns of American history. The history of environmental policy is the history not merely of the Environmental Protection Agency, nor of the public lands, water resource, and wildlife agencies. It is all these things, but it is also the history of policies promoting transportation, industrialization, urbanization and suburbanization, trade, and other uses of environmental resources. It is a product not merely of specialized professions and interest groups, therefore, but of the country's dominant economic and political forces, shaped by elected officials, government agencies, business interests, and citizen demands.

Andrews goes on to explain that "environmental issues are issues not just of science or economics but of governance," and consequently that "environmental policy is not just about managing the environment, but about managing ourselves."[41] The reorientation that he wants to promote here is simple yet profound. Rather than seeking to "deepen" our understanding of environmentalism, he seeks to dramatically expand the breadth of relations that can be appropriately understood as environmental. Yet it is depth of insight that results from an environmental politics so broadly conceived.

Environmental politics in the sense that Andrews presents it *is* our struggle over the creation, use, preservation, alteration, and degradation of place. This struggle is defined by our relationships to these places and our experiences in them, in all their complexity and diversity. This environmentalism politicizes the actions of corporations or landowners in the sense that these are now subject to a degree of public scrutiny that had previously been absent. However, there is certainly no guarantee that such politicization will produce outcomes that are favored by environmental activists, as the "wise use" movement ought to make clear. By understanding and responding to the struggle to define place, however, environmentalists might offer a persuasive response to the arguments of this movement, one that honors its concern with livelihood as a relation to place while challenging its alliance with large resource extraction corporations in the struggle to preserve this relationship.

We have seen that the intention of much critical environmentalist thought is to break out of the single issue box within which ostensibly pluralist political systems and ideologies so often limit environmental concerns. While I have argued that their now-familiar strategy for doing so is not a successful one, their intent is nonetheless important. By putting experience and place at the heart of environmental politics, we can see this politics anew. Not only must environmentalists be compelled to confront the inescapably political dimensions of judgment and interpretation, but political actors and political theorists also ought to be compelled to confront the absolute centrality of the struggle over nature to our policies, politics, and visions of social order.

Notes

Chapter 1

1. Kirkpatrick Sale, *Dwellers in the Land: The Bioregional Vision* (San Francisco: Sierra Club Books, 1985), 49.

2. Bill Devall and George Sessions, *Deep Ecology: Living As If Nature Mattered* (Salt Lake City: Peregrine Smith, 1985), 65.

3. For example, Bill Devall, "Earthday 25: A Retrospective of Reform Environmental Movements," *Philosophy in the Contemporary World* 2 (Winter 1995): 9–15.

4. My goal in this book is neither to explore nor to defend the empirical evidence that has led many to conclude that we face an environmental crisis or crises. Instead, I seek to understand the implications of viewing environmental concerns in this way for our philosophical and political reflections. The best place to begin exploring the evidence is the annual reports issued by the Worldwatch Institute: Lester R. Brown and others, eds., *State of the World*, annual report (New York: W.W. Norton). See also Kevin T. Pickering and Lewis A. Owen, *An Introduction to Global Environmental Issues* (London: Routledge, 1994). For an argument that appeals to crisis are overstated, see Gregg Easterbrook, *A Moment on the Earth: The Coming Age of Environmental Optimism* (New York: Viking, 1995). While Easterbrook's case is well argued on some important points, it is poorly documented.

5. For a thought-provoking sketch of several possible projects for "green political theory" see Andrew Dobson, "Afterword," in *The Politics of Nature: Explorations in Green Political Theory*, Andrew Dobson and Paul Lucardie, eds. (London: Routledge, 1993).

6. Andrew Dobson, *Green Political Thought*, 2nd ed. (London: Routledge, 1995), 24.

7. See Val Plumwood, "Inequality, Ecojustice, and Ecological Rationality," reprinted in *Debating the Earth: The Environmental Politics Reader*, John S. Dryzek and David Schlosberg, eds. (Oxford: Oxford University Press, 1998), 559–583; Robert Goodin, "Enfranchising the Earth, and its Alternatives," *Political Studies* 44 (1996): 835–849.

8. Gus diZerega, "Social Ecology, Deep Ecology, and Liberalism," *Critical Review* 6 (Spring-Summer 1992): 327–328.

9. Paul Wapner, "Politics Beyond the State: Environmental Activism and World Civic Politics," *World Politics* 47(April 1995): 311–340. Cf. Douglas Torgerson, *The Promise of Green Politics: Environmentalism and the Public Sphere* (Durham, N.C.: Duke University Press, 1999).

10. See especially Leo Strauss, *Natural Right and History* (Chicago: University of Chicago Press, 1950). For a discussion specific to environmentalism, see Lewis P. and Sandra K. Hinchman, "'Deep Ecology' and the Revival of Natural Right," *Western Political Quarterly* 42 (Fall 1989): 201–228.

11. *Keywords: A Vocabulary of Culture and Society* (London: Fontana, 1976), 219.

12. Peter Coates, *Nature: Western Attitudes Since Ancient Times* (Berkeley: University of California Press, 1998), 3.

13. Raymond Williams, "Ideas of Nature" in *Problems in Materialism and Culture: Selected Essays* (London: Verso, 1980), 69.

14. Williams, "Ideas of Nature," 70.

15. Coates, *Nature*, 3.

16. Kate Soper, *What Is Nature? Culture, Politics and the Non-Human* (Oxford: Blackwell, 1995), 156.

17. Williams, *Ideas of Nature*, 68–69.

18. For a survey of these conceptions, see J.J. Clarke, *Voices of the Earth: An Anthology of Ideas and Arguments* (New York: George Braziller, 1994).

19. See, for example, Alexandre Koyré, *From the Closed World to the Infinite Universe* (Baltimore, Md.: Johns Hopkins University Press, 1957); R.G. Collingwood, *The Idea of Nature* (Oxford: Oxford University Press, 1945); Carolyn Merchant, *The Death of Nature: Women, Ecology and the Scientific Revolution* (New York: HarperCollins, 1989).

20. Rupert Sheldrake, *The Rebirth of Nature: The Greening of Science and God* (London: Century, 1990), 14.

21. And, of course, an author's claims may or may not correspond to her or his actual intentions, or to the arguments within a work.

22. My approach to interpretation is, on this point, in sympathy with John Dewey's argument that: "what philosophy has been unconsciously, without knowing or intending it, and, so to speak, under cover, it must henceforth be openly and deliberately. When it is acknowledged that under disguise of dealing with ultimate reality, philosophy has been occupied with the precious values embedded in social traditions, that it has sprung from a clash of social ends and from a conflict of inherited institutions with incompatible contemporary tendencies, it will be seen that the task of future philosophy is to clarify men's ideas as to the social and moral strifes of their own day." Dewey, *Reconstruction in Philosophy* (New York: Henry Holt, 1920), 26.

23. Cf. Murray Bookchin, *Remaking Society: Pathways to a Green Future* (Boston: South End Press, 1990); Bob Pepperman Taylor, *Our Limits Transgressed: Environmental Political Thought in America* (Lawrence: University Press of Kansas, 1992); Ariel Kay Salleh, *Ecofeminism as Politics: Nature, Marx and the Postmodern* (London: Zed Books, 1997); John Barry, *Rethinking Green Politics: Nature, Virtue and Progress* (London: Sage, 1999).

24. J. Baird Callicott, "Hume's Is/Ought Dichotomy and the Relation of Ecology to Leopold's Land Ethic," *Environmental Ethics* 4 (Summer 1982): 163–174.

25. See Anthony Weston, "Beyond Intrinsic Value: Pragmatism in Environmental Ethics," *Environmental Ethics* 7 (Winter, 1985): 321–339 for a strong case for a pragmatic approach to environmental ethics on this point. I discuss this further in chapter 6.

26. Andrew Ross, *The Chicago Gangster Theory of Life: Nature's Debt to Society* (London: Verso, 1994), 4.

27. Jane Bennett and William Chaloupka, "Introduction: TV Dinners and the Organic Brunch," in *In the Nature of Things: Language, Politics and the Environment,* Jane Bennett and William Chaloupka, eds. (Minneapolis: University of Minnesota Press, 1993), xvi.

28. One example of unconstructive mean-spiritedness: In *Chicago Gangster Theory of Life,* postmodernist cultural critic Andrew Ross substitutes sarcasm for a critique of writer Bill McKibben. In his book *The Age of Missing Information,* McKibben compared a day in the mountains with a "day" watching (via VCRs) all the TV programs on a large cable system. Unsurprisingly, he found the former to offer important "information" unavailable in the latter. Here is Ross's commentary: "Yikes! For a 32-year-old, McKibben often sounds like Gramps on lithium. This is a man who prefers the mating dance of cranes to semi-naked club kids shaking their Lycra-clad booties on MTV." (179).

29. An excellent analysis of postmodernist discussions of "nature" in relation to environmentalist concerns is in Soper, *What is Nature?*

30. My brief comments here are necessarily qualified by an awareness of the diversity of views identified as postmodern.

31. Timothy W. Luke's recent work, which is penetrating in many respects, nonetheless exemplifies this single-minded focus on critique. See both *Ecocritique: Contesting the Politics of Nature, Economy, and Culture* (Minneapolis: University of Minnesota Press, 1997) and *Capitalism, Democracy, and Ecology: Departing from Marx* (Urbana: University of Illinois Press, 1999). I explore some limitations of this in a review essay entitled "What is Environmental Political Theory?" *Political Theory,* 29 (April 2001): 276–288.

32. Thomas Hobbes, *Leviathan, with selected variants from the Latin edition of 1668,* edited by Edwin Curley (Indianapolis, Ind.: Hackett, 1994), chapter 13, 75.

33. For this claim to make much sense, however, we must go beyond an oversimplified equation of Aristotle's conception of nature with the anthropocentrism expressed in the passage at *Politics* 1256b15–20. There he characterizes animals

and plants as existing wholly for the sake of man. In chapter 5, I sketch a fuller picture of Aristotle's conception of nature from his works on natural philosophy; the result is to displace this passage from center stage.

Chapter 2

1. Arne Naess, "The Shallow and the Deep, Long-Range Ecology Movement: A Summary," *Inquiry* 16 (1973): 95–100.

2. Murray Bookchin, *Remaking Society: Pathways to a Green Future* (Boston: South End Press, 1990).

3. Andrew Dobson, *Green Political Thought,* 2nd ed. (London: Routledge, 1995).

4. Robyn Eckersley, *Environmentalism and Political Theory: Toward an Ecocentric Approach* (Albany: State University of New York Press, 1992).

5. Warwick Fox offers a comprehensive discussion of these dichotomies in *Toward a Transpersonal Ecology: Developing New Foundations for Environmentalism* (Boston: Shambhala, 1990), 22–40. Arne Naess defines an "ecosophy" as "a philosophical worldview or system inspired by conditions of life in the ecosphere" in *Ecology, Community and Lifestyle* (Cambridge: Cambridge University Press, 1989), 38. The language of "paradigms" is also prevalent and generally synonymous with "worldviews." Cf. George Sessions, "Ecological Consciousness and Paradigm Change," in *Deep Ecology,* Michael Tobias, ed. (San Diego: Avant Books, 1985); John Rodman, "Paradigm Change in Political Science: An Ecological Perspective," *American Behavioral Scientist,* 24 (1980): 49–74; Alan R. Drengson, *Beyond Environmental Crisis: From Technocrat to Planetary Person* (New York: Peter Lang, 1989), 42; William Ophuls and A. Stephen Boyan, Jr., *Ecology and the Politics of Scarcity Revisited: The Unraveling of the American Dream* (New York: W.H. Freeman, 1992), 4; Charlene Spretnek and Fritjof Capra, *Green Politics* (Santa Fe, N.M.: Bear, 1986), xix; Lester W. Milbrath, "The World Is Relearning Its Story about How the World Works," in *Environmental Politics in the International Arena,* Sheldon Kamieniecki, ed. (Albany: State University of New York Press, 1993). All these uses of "paradigms" attempt to build upon Kuhn's analysis of the history of natural science in Thomas S. Kuhn, *The Structure of Scientific Revolutions,* 2nd ed. (Chicago: University of Chicago Press, 1970). For helpful discussion, see Richard Routley, "Roles and Limits of Paradigms in Environmental Thought and Action," in *Environmental Philosophy,* Robert Elliot and Arran Gare, eds. (New York: University of Queensland Press, 1983).

6. The terms "worldview," "paradigm," and "communal conception" are used interchangeably in this chapter.

7. On this latter connection, see Anna Bramwell, *Ecology in the 20th Century: A History* (New Haven, Conn.: Yale University Press, 1989); Michael E. Zimmerman, "The Threat of EcoFascism," *Social Theory and Practice* 21 (Summer 1995): 207–238; Robert A. Pois, *National Socialism and the Religion of Nature* (London: Croon Helm, 1986).

8. Naess, *Ecology, Community, Lifestyle,* 40; Murray Bookchin "Recovering Evolution: A Reply to Eckersley and Fox," *Environmental Ethics* 12 (1990): 255.

9. See especially, Bookchin, *Remaking Society,* 7–18 and "Social Ecology Versus Deep Ecology," *Socialist Review* 4 (1988): 9–29.

10. Robyn Eckersley, "Divining Evolution: The Ecological Ethics of Murray Bookchin," *Environmental Ethics* 11 (1989): 99–116.

11. Roderick Nash, "Aldo Leopold's Intellectual Heritage," in *Companion to A Sand County Almanac,* J. Baird Callicott, ed. (Madison: University of Wisconsin Press, 1987), 75; and Wallace Stegner, "The Legacy of Aldo Leopold," in *Companion,* 233.

12. J. Baird Callicott, "The Land Aesthetic," in *Companion,* 157.

13. "The land" is a term that he used as a collective description of the "soils, waters, plants, and animals" that constitute it. Aldo Leopold, *A Sand County Almanac, with Essays on Conservation from Round River* (New York: Ballantine Books, 1970), 239.

14. Stegner, "The Legacy of Aldo Leopold," 234–235.

15. Aldo Leopold, "[1947] Forward," in *Companion,* 281.

16. Leopold, *Sand County,* 258.

17. Leopold, *Sand County,* 259.

18. Leopold, *Sand County,* 260–261.

19. Leopold, *Sand County,* 240.

20. Peter Coates, *Nature: Western Attitudes Since Ancient Times* (Berkeley: University of California Press, 1998), 12.

21. Langdon Winner, "The State of Nature Revisited," in *The Whale and the Reactor* (Chicago: University of Chicago Press, 1986), 131.

22. While White's thesis is famous for its critique of Christianity based upon its ecological effects, the specific links that he articulates between Christianity and modern technology and society are not original. For the argument that our "enlightened" secular age is still very much the product of Christian thought, see Carl Becker, *The Heavenly City of the Eighteenth Century Philosophers* (New Haven, Conn.: Yale University Press, 1959). For the more specific argument that Christianity laid the groundwork for modern science and technology, see Harvey Cox, *The Secular City* (New York: Macmillan, 1965), 22–23; and Johannes Metz, *Theology of the World* (New York: Herder and Herder, 1969). In contrast to White, both present this relationship in a positive light.

23. Lynn White, Jr. "The Historical Roots of Our Ecologic Crisis." *Science* 155 (1967): 1205.

24. See David and Eileen Spring, eds., *Ecology and Religion in History* (New York: Harper and Row, 1974); and Ian Barbour, ed., *Western Man and Environmental Ethics: Attitudes toward Nature and Technology* (Reading, Mass.: Addison-Wesley, 1973).

25. In contrast to the voluntarism that he sees at the heart of the dominant Western versions of Christianity, White characterizes Eastern strands as "intellectu-

alist." Here the notion of acting upon nature was not central ("Historical Roots," 1206). Even within the Western Christian tradition, White points to St. Francis of Assisi as the basis for a minority interpretation (1207). See also Lynn White, Jr., "Continuing the Conversation," in *Western Man and Environmental Ethics: Attitudes toward Nature and Technology,* Ian Barbour, ed. (Reading, Mass.: Addison-Wesley, 1973), 58.

26. White, "Continuing the Conversation," 61.

27. In addition to the works considered here, some of the most recent books include those by John Barry, Andrew Dobson, Tim Hayward, Timothy Luke, David Schlosberg, Avner de-Shalit, and Douglas Torgerson.

28. Eckersley, *Environmentalism and Political Theory,* 28.

29. Eckersley, *Environmentalism and Political Theory,* 49, 51.

30. Eckersley, *Environmentalism and Political Theory,* 59. As noted above, she criticizes Bookchin for failing to heed this injunction in "Divining Evolution."

31. Eckersley, *Environmentalism and Political Theory,* 52.

32. Eckersley, *Environmentalism and Political Theory,* 86. See also page 131.

33. For a helpful discussion of the differences between convergent and divergent interpretations of environmentalism and green political thought, see Dobson, *Green Political Thought,* 199–210.

34. The approaches explored are orthodox Marxism, critical theory, ecosocialism, ecoanarchism, and ecocommunalism.

35. Eckersley, *Environmentalism and Political Theory,* 185.

36. Eckersley, *Environmentalism and Political Theory,* 28.

37. See sources in footnote 9.

38. Murray Bookchin, *The Ecology of Freedom: The Emergence and Dissolution of Hierarchy* (Palo Alto, Calif.: Cheshire Books, 1982), 1. Also *Remaking Society,* 44–46. This thesis may appear familiar to readers of Horkheimer, Adorno, and Marcuse. However, Bookchin seeks to reverse the causal arrows in the critical theorists' argument. The latter present the domination of nature as the reflection of instrumental reason and a precursor to human domination. See *Ecology of Freedom,* 283–284 n.

39. Bookchin, *Ecology of Freedom,* 18.

40. Bookchin, *Ecology of Freedom,* 3.

41. See, for example, Murray Bookchin, review of *Ecology as Politics,* by André Gorz, in *Telos* 46 (1980–81): 176–190.

42. Bookchin, *Ecology of Freedom,* 278. See also "Thinking Ecologically," in *The Philosophy of Social Ecology: Essays on Dialectical Naturalism* (Montreal: Black Rose Books, 1990), 176; and the exchange between Bookchin and Eckersley on this issue in Bookchin, "Recovering Evolution," and Eckersley, "Divining Evolution."

43. Bookchin, "Introduction: A Philosophical Naturalism," in *Social Ecology,* 20.

44. Bookchin, *Ecology of Freedom*, 32.

45. Bookchin, *Ecology of Freedom*, 289.

46. Bookchin, "Philosophical Naturalism," 42–45.

47. Bookchin, "Freedom and Necessity in Nature," in *Social Ecology*, 115.

48. Bookchin, "Philosophical Naturalism," 43; "Thinking Ecologically," 176.

49. Bookchin, "Thinking Ecologically," 176–177.

50. Bookchin, "Thinking Ecologically," 178. For a detailed treatment, see Bookchin, *Ecology of Freedom*, chapters 2–5.

51. Bookchin, "Philosophical Naturalism," 45.

52. Bookchin, *Ecology of Freedom*, 318.

53. Bookchin, *Ecology of Freedom*, 274; 289.

54. Bookchin, "Recovering Evolution," 263.

Chapter 3

1. See especially, John Barry, "The Limits of the Shallow and the Deep: Green Politics, Philosophy, and Praxis," *Environmental Politics* 3 (1994): 369–394, Bob Pepperman Taylor, "Environmental Ethics and Political Theory," *Polity* 23 (Summer, 1991): 567–583.

2. Val Plumwood, *Feminism and the Mastery of Nature* (London: Routledge, 1993), 70, emphasis added.

3. Peter Marshall, *Nature's Web: An Exploration of Ecological Thinking* (London: Simon and Schuster, 1992), 5.

4. Fritjof Capra identifies Descartes as the primary villain in *The Turning Point: Science, Society and the Rising Culture* (New York: Bantam Books, 1982). A variety of others focus on contractarianism: see Andrew Brennan, *Thinking About Nature: An Investigation of Nature, Value and Ecology* (Athens: University of Georgia Press, 1988), chapters 11, 12; Michel Serres, *The Natural Contract*, translated by Elizabeth MacArthur and William Paulson (Ann Arbor: University of Michigan Press, 1995); and William Ophuls, *Ecology and the Politics of Scarcity: Prologue to a Political Theory of the Steady State* (San Francisco: W.H. Freeman, 1977), 215. Bruno Latour offers a critical discussion of "modernity" as identified with dualism in *We Have Never Been Modern*, translated by Catherine Porter (Cambridge, Mass.: Harvard University Press, 1993).

5. Brennan, *Thinking About Nature*, 185.

6. Plumwood, *Feminism and the Mastery of Nature*, 71.

7. Plumwood, *Feminism and the Mastery of Nature*, 43.

8. Plumwood and other ecofeminists quite explicitly draw their environmentalist critique of dualism from the insights of (other) feminist theorists. Yet dualism plays a more limited role within much (non-eco-)feminist theory than it does

within the environmentalist theories under consideration here. Feminist discussions of dualism are commonly an analysis of gendered distinctions *within human culture*. The distinction analyzed is typically between a realm of *necessity* and a realm defined as freedom, politics, or culture. The typical claim is that Western theorists speak of women's roles *as if* they were part of nature, and apart from (male) culture. This sort of feminist critique need not conclude that Western social and political thought emerges *apart* from any conception of nature, but only that *within* Western tradition and culture, inegalitarian gender divisions have been rationalized by the proclaimed affinity between women and nonhuman nature.

By contrast, ecofeminists such as Plumwood advance an argument that encompasses the former claim as well as the latter. Certainly, much can be learned by studying the overlapping justifications for the subjugation of women and nonhuman nature, a project initiated and led by the work of ecofeminist researchers. The argument that a human–nature dualism lies at the root of Western thought, however, is a mistaken overextension of this project. In the following section of my chapter, I consider the work of another prominent ecofeminist thinker, Carolyn Merchant, whose analysis leads her to overextend the insights of her historical analysis in the opposite direction.

9. For example, "Some Sources of the Deep Ecology Perspective," in Bill Devall and George Sessions, *Deep Ecology: Living As If Nature Mattered* (Salt Lake City: Peregrine Smith, 1985).

10. Devall and Sessions, *Deep Ecology,* chapter 2.

11. In the following section of this chapter, I discuss environmentalist interpretations of both Aristotle and Thomas Hobbes that characterize them as exemplars of Western thought's pattern of deriving politics from nature. Yet surely philosophers as central to Western thought as these cannot be plausibly categorized as part of such a "minority tradition."

12. Though still far from unequivocal or universal. Eugene C. Hargrove in *Foundations of Environmental Ethics* (Englewood Cliffs, N.J.: Prentice Hall, 1989) seeks to trace a thread of aesthetic and natural history thought within the Western tradition (going back to the Romantics) that embraces *and values* nature and the natural. Hargrove contrasts this thread with the more explicitly philosophical thread of Western thought.

13. J.J. Clarke provides a survey of various conceptions in *Voices of the Earth: An Anthology of Ideas and Arguments* (New York: George Braziller, 1994), while Clarence Glacken offers a more comprehensive discussion of them in *Traces on the Rhodian Shore: Nature and Culture in Western Thought from Ancient Times to the End of the Eighteenth Century* (Berkeley: University of California Press, 1967).

14. For discussions of both, see R.G. Collingwood, *The Idea of Nature* (Oxford: Oxford University Press, 1945); Alexandre Koyré, *From the Closed World to the Infinite Universe* (Baltimore, Md.: Johns Hopkins University Press, 1957).

15. *Politics* 1256b15. Despite frequent characterization of it as at the core of his conception of nature, this unequivocally anthropocentric statement from the

Politics is not central to—and probably not even a part of—Aristotle's elaborately developed teleological conception of nature, a point I develop in chapter 5.

16. *Politics* 1253a1.

17. Hobbes explicitly criticized Descartes on this score. See [Thomas Hobbes and] René Descartes, "The Third Set of Objections with the Author's Reply," in *Philosophical Works of Descartes,* edited and translated by Elizabeth S. Haldane and G.R.T. Ross (Cambridge: Cambridge University Press, 1911); and Frithiof Brandt, *Thomas Hobbes' Mechanical Conception of Nature* (Copenhagen and London: Levin & Munksgaard and Hachette, 1928).

18. There is real interpretive tension between Merchant's "organic Aristotle" and the "anthropocentric Aristotle" sketched in the earlier paragraph. Each is the result of a consideration of only a portion of Aristotle's writings as the basis for conclusions about his conception of nature. Common to both, however, is the sense that politics was derived from a conception of nature.

19. Carolyn Merchant, *The Death of Nature: Women, Ecology, and the Scientific Revolution* (New York: HarperCollins, 1989), 193.

20. Merchant, *Death of Nature,* 206.

21. Freya Mathews, *The Ecological Self* (Savage, Md.: Barnes and Noble, 1991), 29. Of course, Hobbes's work preceded Newton. Mathews argues that the fundamental characteristics of the view that she labels Newtonian, however, were already developed during Hobbes's career.

22. Mathews, *Ecological Self,* 29.

23. Mathews, *Ecological Self,* 29.

24. See also Capra, *The Turning Point,* 75–97, 265–419; and J. Baird Callicott, "Intrinsic Value, Quantum Theory and Environmental Ethics," *Environmental Ethics* 7 (1985): 257–275.

25. Thomas A. Spragens, Jr., in *The Politics of Motion: The World of Thomas Hobbes* (Lexington: University Press of Kentucky, 1973) offers an especially clear account along these lines, in which Hobbes is interpreted as transforming the content of the Aristotelian "paradigm," while nonetheless adhering to the same form of theorizing in which political thought is derived from an account of the nature of motion.

26. Mathews, *Ecological Self,* 29.

27. Ted Benton, *Natural Relations: Ecology, Animal Rights and Social Justice* (London: Verso, 1993), 175.

28. This insight has been central for some extremely interesting and insightful work by environmental historians. See, for example, William Cronon, *Nature's Metropolis: Chicago and the Great West* (New York: W.W. Norton, 1991), and Donald Worster, ed., *The Ends of the Earth: Perspectives on Modern Environmental History* (Cambridge: Cambridge University Press, 1988).

29. Michael J. Sandel, "The Procedural Republic and the Unencumbered Self," *Political Theory* 12 (February 1984): 90–91; also Alasdair MacIntyre, *After Virtue,* 2nd ed. (Notre Dame: University of Notre Dame Press, 1984).

30. The breathtaking rapidity of contemporary developments in biotechnology certainly make this less difficult to imagine. For a discussion, see John M. Meyer, "Rights to Life?: On Nature, Property and Biotechnology," *Journal of Political Philosophy* 8 (June 2000): 154–175.

31. Of course, most environmentalist thinkers rightly deny this, although a few notorious examples of explicit misanthropy are frequently recycled by critics. Nonetheless, at the conceptual level of my inquiry here, there does seem to be some reason for critics to be concerned that by rejecting a dualism in which the privileged half is closely identified with humanism and artifice, some environmentalists may inadvertently reject the positive values associated with these as well. Luc Ferry, in *The New Ecological Order,* translated by Carol Volk (Chicago: University of Chicago Press, 1995), presents an important yet highly problematic example of such an argument.

Chapter 4

1. Sheldon S. Wolin, *Hobbes and the Epic Tradition of Political Theory* (Los Angeles: W. A. Clark Memorial Library, University of California at Los Angeles, 1970), 19.

2. For an excellent critique of the misuses and overextension of the concept of modernity, see Bernard Yack, *The Fetishism of Modernities: Epochal Self-Consciousness in Contemporary Social and Political Thought* (Notre Dame, Ind.: University of Notre Dame Press, 1997).

3. Thomas Hobbes, *Elements of Philosophy, the first section, concerning Body,* in *The English Works of Thomas Hobbes,* Sir William Molesworth, ed. Vol. 1 (London: 1839), Epistle Dedicatory, viii. This is the English translation of Hobbes's *De Corpore* published during his lifetime. Hereafter cited as *De Corpore.*

4. *De Corpore,* Epistle Dedicatory, viii. I use the terms "natural science" and "natural philosophy" interchangeably.

5. *De Corpore,* chapter 8, #19, 115–116.

6. *De Corpore,* chapter 10, #7, 131. See also Thomas A. Spragens, Jr., *The Politics of Motion: The World of Thomas Hobbes* (Lexington: University Press of Kentucky, 1973), chapter 2.

7. While not denying the significance of matter, and hence Hobbes's "materialism," Frithiof Brandt, one of the few careful students of his natural philosophy, has convincingly argued that his conception of nature is best described as "motionalist." *Thomas Hobbes' Mechanical Conception of Nature* (Copenhagen and London: Levin & Munksgaard and Hachette, 1928), 379. See also Robert Hugh Kargon, *Atomism in England from Hariot to Newton* (Oxford: Clarendon Press, 1966), 57–58.

8. See especially, Hobbes's exchange with Bishop Bramhall in *The Questions Concerning Liberty, Necessity and Chance* in *English Works,* Vol. 5, for an explication of his determinism.

9. Thomas Hobbes, *Leviathan, with selected variants from the Latin edition of 1668,* edited with an introduction by Edwin Curley (Indianapolis, Ind.: Hackett, 1994), chapter 46, 459.

10. John W. N. Watkins, *Hobbes's System of Ideas,* 2nd ed. (Aldershot, England: Gower, 1989), 24.

11. *Human Nature and De Corpore Politico,* edited with an introduction by J.C.A. Gaskin (Oxford: Oxford University Press, 1994), chapter 11, 64. These two titles compose the work more commonly known (and hereafter cited) as *The Elements of Law.* See also *Leviathan,* chapter 11, 62.

12. Alexandre Koyré, "The Significance of the Newtonian Synthesis," in *Newtonian Studies* (Cambridge, Mass.: Harvard University Press, 1965), 24.

13. Leo Strauss argues that the opposition to anthropomorphism prevents Hobbes from developing his political upon his natural philosophy *at all.* This contention serves as the premise for Strauss's search for an alternative, humanistic, basis for Hobbes's politics. See *Political Philosophy of Hobbes: Its Basis and Its Genesis,* translated by Elsa Sinclair (Chicago: University of Chicago Press, 1952), xiii.

14. Richard Peters, *Hobbes* (Harmondsworth, U.K.: Penguin, 1956), 91–92, and Watkins, *Hobbes's System of Ideas,* 87–88. For an especially thorough discussion of endeavour/*conatus* in Hobbes, which nonetheless fails to recognize its bridging function between the physical and the psychical, see Brandt, *Mechanical Conception,* 295–315.

15. *De Corpore,* chapter 15, #2, 206. See also *Elements of Law,* chapter 7, #2, 43–44; *Leviathan,* chapter 6, 28.

16. *De Corpore,* chapter 6, #6, 72. See also *Leviathan,* chapter 1, 7.

17. *Leviathan,* chapter 14.

18. *Leviathan,* chapter 15, 100. Since "law," properly speaking, is understood by Hobbes as a command issued by a sovereign, these "dictates of reason" only truly become law after the creation of sovereignty—and/or if they are recognized as the word of God.

19. On this point, see Edwin Arthur Burtt, *The Metaphysical Foundations of Modern Physical Science,* rev. ed. (London: Routledge & Kegan Paul, 1932), passim, and Alexandre Koyré, "Galileo and Plato," in *Metaphysics and Measurement: Essays in Scientific Revolution* (Cambridge, Mass.: Harvard University Press, 1968), 37.

20. *Elements of Law,* chapter 2, #10, 26. Also *De Corpore,* chapter 8, 101–119; *Leviathan,* chapter 1, 7.

21. *De Corpore,* chapter 2, #9, 20. See also *Leviathan,* chapter 4, 17.

22. *De Corpore,* chapter 8, #2, 103.

23. *Leviathan,* chapter 1, 7. *Elements of Law,* chapter 2, 26. Hobbes engaged in a bitter dispute with Descartes over the question of which of the two developed this doctrine first. See Brandt, *Mechanical Conception,* 135–142.

24. Martin Bertman, *Hobbes: The Natural and the Artifacted Good* (Bern: Peter Lang, 1981), 20. See also Richard Tuck, "Optics and Sceptics: The Philosophical

Foundations of Hobbes's Political Thought," in *Conscience and Casuistry in Early Modern Europe,* Edmund Leites, ed. (Cambridge, Mass.: Cambridge University Press, 1988), 243. Writing about Gassendi, whose argument about nature and perception parallels Hobbes on this point, Tuck concludes: "Accordingly, the only appropriate philosophy had to be a version of nominalism."

25. Hobbes's claim here seems to be that the very existence of our sensations offers the necessary proof of an external nature composed of matter in motion that must serve as their cause. See Tuck, "Optics and Sceptics," 251–254.

26. *Leviathan,* chapter 13, 76.

27. I use the phrases "state of nature" and "natural condition" interchangeably here, as Hobbes seems to do. It may be worth noting, however, that while *Leviathan* offers the most sophisticated development of this concept for Hobbes (chapter 13), the more famous phrase "state of nature" does not occur here. The phrase can be primarily found in *De Cive,* while also occurring in *Elements of Law.*

28. *Leviathan,* chapter 13, 76.

29. *Leviathan,* chapter 17, 109.

30. *Leviathan,* chapter 17, 109. I develop the significance of this point further in the following sections.

31. *Leviathan,* chapter 19, 120.

32. *Leviathan,* chapter 19, 120.

33. *Leviathan,* chapter 15, 100. I develop the significance of this point later.

34. Many nonenvironmentalist scholars of Hobbes's philosophy have offered similar accounts. See Tom Sorell, "The Science in Hobbes's Politics," in *Perspectives on Thomas Hobbes,* G. A. J. Rogers and Alan Ryan, eds. (Oxford: Clarendon Press, 1988), 69, for a characterization of what he terms this "standard account" of Hobbes's philosophy. Cf. Watkins, *Hobbes's System of Ideas;* Peters, *Hobbes;* Spragens, *Politics of Motion;* M. M. Goldsmith, *Hobbes's Science of Politics* (New York: Columbia University Press, 1966).

35. William Ophuls, *Requiem for Modern Politics: The Tragedy of the Enlightenment and the Challenge of the New Millennium* (Boulder, Col.: Westview Press, 1997), 187–188.

36. Thomas Hobbes, *De Cive, also known as Philosophical Rudiments Concerning Government and Society* in *Man and Citizen,* edited with an introduction by Bernard Gert (Indianapolis, Ind.: Hackett, 1991), author's preface to reader, 102–103. This is the English translation published during Hobbes's lifetime, which had been traditionally (though perhaps incorrectly) attributed to him. Richard Tuck's more careful translation of the original Latin from these pages, along with his additional research, significantly strengthens the claim that Hobbes developed and viewed his philosophy as an interdependent whole. Tuck argues convincingly that Hobbes had already generated his whole philosophical system in a relatively sophisticated form before the time of *De Cive*'s first publication in 1642. See Tuck, "Hobbes and Descartes" in *Perspectives on Thomas Hobbes,* 19–20.

37. *De Corpore,* chapter 6, #17, 87–88.

38. The same could be said for Hobbes's first philosophical work to be circulated (but not published), *The Elements of Law.*

39. *De Cive,* author's preface to the reader, 103. As he also emphasizes here, he rushed to publish *De Cive* because he sought to advance his political philosophy as a solution to the civil war then raging in England.

40. *De Corpore,* chapter 6, #17, 73–75. See also chapter 25, #1, 387–388 for this argument advanced in more general terms.

41. See also Watkins, *Hobbes's System of Ideas,* 4–5.

42. *De Corpore,* Epistle Dedicatory, ix.

43. Martin A. Bertman makes a similar point, although his emphasis is on Hobbes's methodological claim to distinctiveness. See *Hobbes: The Natural and the Artifacted Good,* 77–80.

44. *The Assayer* in *Discoveries and Opinions of Galileo,* translated with introduction and notes by Stillman Drake (New York: Anchor, 1957), 237–238.

45. *Leviathan,* chapter 46, 456.

46. *Leviathan,* chapter 13, 78.

47. Thomas A. Spragens, Jr., "The Politics of Inertia and Gravitation: The Functions of Exemplar Paradigms in Social Thought," *Polity* (1973): 300.

48. *De Corpore,* chapter 1, #6, 7.

49. *De Corpore,* chapter 1, #7, 7.

50. *Leviathan,* chapter 13, 76. See also *De Corpore,* chapter 1, #7, 8.

51. *Leviathan,* Introduction, 3.

52. Norberto Bobbio, *Thomas Hobbes and the Natural Law Tradition,* translated by Daniella Gobetti (Chicago: University of Chicago Press, 1993), xi. See also *Leviathan,* Epistle Dedicatory, 1, where Hobbes writes: "in a way beset with those that contend, on one side for too great liberty, and on the other side for too much authority, 'tis hard to pass between the points of both unwounded."

53. *Leviathan,* chapter 30, 229.

54. *Leviathan,* chapter 14, 80.

55. Mary Dietz has argued that Hobbes should be understood as promoting a kind of "civic virtue" among his subjects; a view that seems to challenge my conclusion here about the sorts of state activities ruled out by Hobbes's theory. Yet even Dietz emphasizes that the educative role of the sovereign is quite limited, in that the civic virtue to be inculcated is only that "compatible with the pursuit of civil peace" (96). See "Hobbes's Subject as Citizen," in *Thomas Hobbes and Political Theory,* Mary G. Dietz, ed. (Lawrence: University Press of Kansas, 1990).

56. *Leviathan,* chapter 26, 187.

57. Such motivations are especially noted in his history of the English Civil War. See Thomas Hobbes, *Behemoth: or, The Long Parliament,* Ferdinand Tönnies, ed., with an introduction by Stephen Holmes (Chicago: University of Chicago Press, 1990).

58. Death (especially violent death), so far as we can know, is the *summum malum* according to Hobbes. Only it can end the "perpetual and restless desire of power after power" that Hobbes posits as a "general inclination of all mankind." *Leviathan,* chapter 11, 58.

59. *Leviathan,* chapter 6, 34.

60. Leo Strauss, "On the Basis of Hobbes's Political Philosophy," in *What is Political Philosophy?* (Chicago: University of Chicago Press, 1988), 177.

61. Strauss, "On the Basis of Hobbes's Political Philosophy," 177.

62. Strauss, *Political Philosophy of Hobbes,* 168.

63. Note how this characterization of modern thought as distinctively dualistic also contrasts with the characterization of modern thought as the distinctive product of a modern conception of (mechanistic) nature.

64. Michael Oakeshott, "Introduction to *Leviathan*" reprinted in *Rationalism in Politics and Other Essays,* new and expanded ed. (Indianapolis, Ind.: Liberty Press, 1991), 227 and even more pointedly, "Logos and Telos" in the same volume. Patrick Riley, *Will and Political Legitimacy* (Cambridge, Mass.: Harvard University Press, 1982), 44, highlights the one-sidedness of this point in Oakeshott's interpretation of Hobbes.

65. As a point of comparison, recall Lynn White's indictment of "voluntarism" as a cause of ecological crisis.

66. The subtitle to his 1936 [1952] book on Hobbes.

67. *Aubrey's Life of Hobbes,* reprinted in *Leviathan,* Curley, ed., p. lxvi.

68. George Croom Robertson, *Hobbes* (Edinburgh: Blackwood, 1886), 57.

69. All of this is also complicated by the contested dating and authorship of a text first identified by Ferdinand Tönnies, and usually dated to 1631, titled (by Tönnies) "A Short Tract on First Principles." This text sketches a natural philosophy with parallels to Hobbes's later writings. While Strauss essentially ignored this text (see p. xii of Strauss, *Political Philosophy of Hobbes* for his unsubstantiated dismissal of its significance for his project), in *Hobbes's System of Ideas,* Watkins uses it in a central role in his more derivative account of Hobbes's philosophy.

70. Thomas Hobbes, *Three Discourses: A Critical Modern Edition of Newly Identified Work of the Young Hobbes,* edited by Noel B. Reynolds and Arlene W. Saxonhouse (Chicago: University of Chicago Press, 1995).

71. Strauss, *Political Philosophy of Hobbes,* xii, n1.

72. See "Proof of Authorship and Statistical Wordprint Analysis," in Hobbes, *Three Discourses,* 10–19.

73. Strauss makes much of the latter possibility. If correct, it justifies his esoteric strategy of interpreting Hobbes. It must be said, though, that if Hobbes's intention was to make his published political writings palatable (most especially *Leviathan,* which he published in English in order to attract a wider readership) by submerging the controversial truths that Strauss claims to have identified so that they could be discovered only by a few, he was a distinct failure. For

the controversy over Hobbes's writing in his own time, see John Bowle, *Hobbes and His Critics: A Study in Seventeenth Century Constitutionalism* (Oxford: Jonathan Cape, 1951), and Samuel I. Mintz, *The Hunting of Leviathan: Seventeenth Century Reaction to the Materialism and Moral Philosophy of Thomas Hobbes* (Cambridge: Cambridge University Press, 1962). For an elaboration and defense of the strategy of reading texts esoterically rather than exoterically, see Leo Strauss, *Persecution and the Art of Writing* (Glencoe, Ill.: Free Press, 1952).

74. And, I would note, many first-time readers of *Leviathan* even come away with the sense that Hobbes views humans as "by nature evil," the core of the teaching that Strauss believes Hobbes sought to obscure through his mechanistic science. See Strauss, *Political Philosophy of Hobbes*, 13. Ironically, to the extent that this familiar undergraduate reading of Hobbes comes close to elements of Strauss's supposedly esoteric reading, it is precisely because of the unwitting acceptance of Strauss's argument that, "if we want to do justice to the life which vibrates in Hobbes's political teaching, we must understand that teaching by itself, and not in the light of his natural science." Strauss, "On the Basis of Hobbes's Political Philosophy," 179.

75. *Leviathan*, chapter 6, 33.

76. *Leviathan*, chapter 40, 317.

77. For a helpful discussion of Hobbes's conception of will that argues for its inadequacy as a basis for a polity legitimated by its voluntary creation, see Riley, *Will and Political Legitimacy*, 23–60.

78. Hobbes's contribution to the development of the idea of modern science is usually given little more than a footnote in general histories of the subject. An exception is Robert H. Kargon, who states that "Hobbes was one of the three most important mechanical philosophers of the mid-seventeenth century, along with Descartes and Gassend." *Atomism in England from Hariot to Newton* (Oxford: Clarendon Press, 1966), 54. It seems unquestionable that Hobbes's own seriousness of purpose in applying himself to this subject was at least comparable to that with which he applied himself to the development of his political philosophy. For discussions of some reasons offered for Hobbes's neglect, and an effort at correction, see Steven Shapin and Simon Schaffer, *Leviathan and the Air-Pump: Hobbes, Boyle, and the Experimental Life* (Princeton, N.J.: Princeton University Press, 1985), 7–9 and passim; also Noel Malcolm, "Hobbes and the Royal Society," in *Perspectives on Thomas Hobbes*.

79. *Leviathan*, Introduction, 3.

80. *Leviathan*, chapter 13, 75.

81. *Leviathan*, chapter 13, 76.

82. However, see his description of American Indian tribes as living in the state of nature for a deviation from this emphasis on the hypothetical character of the natural condition. *Leviathan*, chapter 13, 77. See also *Leviathan*, chapter 17, 106.

83. It may be worth noting that the analysis I advance here is also distinct from the critique of Hobbes's state of nature advanced by theorists including

MacPherson and Rousseau. They argue that Hobbes has merely imagined the removal of historically extant laws and institutions, while retaining their socializing effects upon individuals, in his characterization of the state of nature. By contrast, I wish to emphasize that Hobbes's conception is defined as a negation of his own ideal of sovereignty. Cf. C.B. MacPherson, *The Political Theory of Possessive Individualism* (Oxford: Oxford University Press, 1962), 17–46; Jean-Jacques Rousseau, *Discourse on the Origin and Foundations of Inequality,* in *The First and Second Discourses,* edited by Roger D. Masters and translated by Roger D. and Judith R. Masters (New York: St. Martin's Press, 1964).

84. *Leviathan,* chapter 13, 78. See also François Tricaud, "Hobbes's Conception of the State of Nature from 1640 to 1651: Evolution and Ambiguities," in *Perspectives on Thomas Hobbes,* 122. Tricaud (like Leo Strauss but without the same normative argument) observes that this formulation of the state of nature represents an evolution in Hobbes's presentation. He argues convincingly that earlier, in *The Elements of Law,* Hobbes's account was portrayed in more moralistic terms as an outcome of human passions. The shift to a more completely structural account by the time of the publication of *Leviathan* offers an account more consistent with the distinctions between nature and the state of nature that I develop here.

85. *Elements of Law,* chapter 15, #1, 82.

86. On Hobbes's opposition to experimental science, and his view that it represented a potentially threatening, independent realm of power, see Shapin and Schaffer, *Leviathan and the Air-Pump,* 327.

87. *Leviathan,* chapter 15, 100.

88. *De Cive,* chapter 3, #31, 150.

89. See Tricaud, "State of Nature," 113–123, for a careful discussion of these two and other causes of the state of war described by Hobbes.

90. Tricaud, "State of Nature," 114.

91. Thomas Spragens, in *The Politics of Motion,* argues for a close link between Hobbes's natural and political philosophy, yet characterizes this link as analogical. In his conclusion, he nonetheless acknowledges that "all sorts of intermediate and contingent judgments . . . intervene" between the two (205). If so, and I believe he is correct here, then a relationship characterized as an analogy is probably not one that is nearly as tightly linked as Spragens suggests elsewhere in this book.

92. *Leviathan,* chapter 13, 78.

93. This understanding of natural law as the prescriptions derived from reasoning in a state of nature is at odds with at least one important school of Hobbes scholarship that views his natural law as something of a deontology. See especially, Howard Warrender, *The Political Philosophy of Hobbes: His Theory of Obligation* (Oxford: Clarendon Press, 1957). If Warrender is right about Hobbes's project, however, then all of Hobbes's efforts to forge links between natural and political philosophy must be dismissed as aberrations.

94. *Leviathan,* chapter 13, 78.

95. See Deborah Baumgold, "Hobbes's Political Sensibility: The Menace of Political Ambition," in *Thomas Hobbes and Political Theory.*

Chapter 5

1. *Leviathan,* chapter 46, 457.

2. *Generation of Animals* 744b17.

3. For discussions of the contemporary science of ecology that emphasize its distancing from the "balancing" concept, see Daniel B. Botkin, *Discordant Harmonies: A New Ecology for the Twenty-First Century* (New York: Oxford University Press, 1990) and Donald Worster, "The Ecology of Order and Chaos," in *The Wealth of Nature* (Oxford: Oxford University Press, 1993).

4. Donald Worster identifies biologist Ernest Haeckel as the first to use the term "ecology" in the 1860s. *Nature's Economy: A History of Ecological Ideas* (Cambridge: Cambridge University Press, 1985), 192. On the resonance of Aristotle's conception of nature with contemporary environmentalist conceptions, see Worster's introduction and p. 37; Laura Westra, *An Environmental Proposal for Ethics: The Principle of Integrity* (Lanham, Md.: Rowman and Littlefield, 1994), 134–142; and Laura Westra and Thomas M. Robinson, eds., *The Greeks and the Environment* (Lanham, Md.: Rowman and Littlefield, 1997), Part III.

5. Paul W. Taylor, *Respect for Nature: A Theory of Environmental Ethics* (Princeton, N.J.: Princeton University Press, 1986), 119–129.

6. Val Plumwood, *Feminism and the Mastery of Nature* (London: Routledge, 1993), 135. Plumwood, however, rejects Aristotle's teleology as irredeemably anthropocentric.

7. *Politics,* 1256b15. See Eugene C. Hargrove, *Foundations of Environmental Ethics* (Englewood Cliffs, N.J.: Prentice Hall, 1989); J.D. Hughes, "Ecology in Ancient Greece," *Inquiry* 18 (1975): 122; Plumwood, *Feminism and the Mastery of Nature,* 46; Peter Marshall, *Nature's Web: An Exploration of Ecological Thinking* (London: Simon and Schuster, 1992), 75.

8. For an extended argument that presents Aristotle as the exemplar of a derivative relationship between nature and politics who established the model or "paradigm" to which Hobbes adheres, see Thomas A. Spragens, Jr., *The Politics of Motion: The World of Thomas Hobbes* (Lexington: University Press of Kentucky, 1973).

9. The conflation of "nature" and "necessity" in this reading of Aristotle is significant. Hannah Arendt, for example, asserts that "Natural community in the household therefore was born of necessity, and necessity ruled over all activities performed in it" (Hannah Arendt, *The Human Condition,* Chicago: University of Chicago Press, 1958, 30). Or later, writing about the emergence of a "social realm" in which the "life process" has entered the "public domain," she characterizes this as an "unnatural growth of the natural" (*Human Condition,* 47). As I discuss in subsequent sections of this chapter, however, this conflation of nature and necessity is not true to Aristotle's own efforts to define these concepts.

10. Wendy Brown, *Manhood and Politics: A Feminist Reading in Political Theory* (Lanham, Md.: Rowman and Littlefield, 1988), 26.

11. Arendt, *Human Condition*, 29–37.

12. Brown, *Manhood and Politics*, 32.

13. *Physics* 192b8–15.

14. *Physics* 192b15.

15. *On the Heavens* 301b17. Sarah Waterlow argues that this definition of nature is the "fundamental idea" behind all of Aristotle's physical doctrines. See *Nature, Change, and Agency in Aristotle's* Physics: *A Philosophical Study* (Oxford: Clarendon Press, 1982), 1 and passim; also Frederick J.E. Woodbridge, *Aristotle's Vision of Nature* (New York: Columbia University Press, 1965), 52–53.

16. *Physics* 192b19.

17. See *Physics* 194b24–35; *Metaphysics* 1013a24–b4.

18. *Parts of Animals* 639b15; see also *Physics* 195a20–25.

19. *Physics* 194a30.

20. *Progression of Animals* 704b12–15.

21. *Physics* 194a34; *Progression of Animals* 704b12–15.

22. See especially, *Physics* Book II chapter 9.

23. *Physics* 198b35; also *Eudemian Ethics* 1247a32, *On the Heavens* 301a.

24. *Physics* 199b15–20.

25. W. Wieland emphasizes this point. See "The Problem of Teleology," in *Articles on Aristotle,* Vol. 1, Jonathan Barnes, Malcolm Schofield, and Richard Sorabji, eds. (London: Duckworth: 1975), 159 and passim.

26. *Nicomachean Ethics* 1140a10. See also his critique of Democritus for mistakenly conflating necessity with nature, *Generation of Animals* 789b3–5.

27. He is explicit in distancing himself from such a view in *Physics* 199a21.

28. See Wieland, "The Problem of Teleology," 153–159.

29. *Progression of Animals* 706a19.

30. Cary J. Nederman, "The Puzzle of the Political Animal: Nature and Artifice in Aristotle's Political Theory," *Review of Politics* 56 (Spring 1994): 290. See also Arlene Saxonhouse, "Aristotle: Defective Males, Hierarchy and the Limits of Politics," in *Women in the History of Political Thought: Ancient Greece to Machiavelli* (New York: Praeger, 1985) for a similar interpretation on this point.

31. W. D. Ross, *Aristotle,* 5th ed. (London: Methuen, 1949), 68.

32. *Physics* 212b29.

33. *Physics* 193b22–194a5.

34. See Joe Sachs, *Aristotle's* Physics: *A Guided Study* (New Brunswick, N.J.: Rutgers University Press, 1995), 56–57. It is worth noting that Aristotle's emphasis on the importance of context or place precludes many assumptions necessary for the establishment of a natural science reliant upon experiment and experimen-

tal manipulation. If a body's nature is exhibited only or best in its natural context, then the forcible control of conditions necessary for experiment will at best offer no additional insights to a study of the body in its natural context, or at worst will prevent the experimenter from discovering a body's true nature, since this is no longer revealed in the absence of its natural place and conditions.

35. Alexandre Koyré, "Galileo and Plato," in *Metaphysics and Measurement: Essays in Scientific Revolution* (Cambridge, Mass.: Harvard University Press, 1968), 24.

36. *Politics* 1253a2–4.

37. *Nicomachean Ethics* 1094a26–28; also *Politics* 1282b16; *Eudemian Ethics* 1218b10–14.

38. *Nicomachean Ethics* 1094b5–8.

39. *Nicomachean Ethics* 1094b7.

40. Of course, my argument here would be moot if in fact no sort of differentiation did exist in Aristotle's time. This seems to be the argument of Steven Taylor Holmes in "Aristippus in and out of Athens," *American Political Science Review* 73 (1979): 113–138. But arguments such as Holmes's overlook the clarification that I seek to offer here. Perhaps it was the case that the economy, the household, education, etc., were all in fact ordered by politics in Aristotle's time (although this point itself is highly questionable). It seems nonetheless evident from Aristotle's own writing that these forms of activity *were* recognized as identifiable and distinct from political activity; otherwise it would be wholly unnecessary and even incomprehensible to argue for a conception of politics as the *"master* science," since no identifiable, *subsidiary* "sciences" would be recognized for it to direct.

41. Keyt and Miller, "Introduction" to *A Companion to Aristotle's* Politics, David Keyt and Fred D. Miller, Jr., eds. (Cambridge, Mass.: Blackwell, 1991), 2. See also Ernest Barker, "Introduction" to *The Politics of Aristotle* (Oxford: Oxford University Press, 1946), lxiii–lxvii.

42. See his discussions of Persia (*Politics* 1326b2–7) and Egypt (*Politics* 1329b33). Also see C.C.W. Taylor, "Politics," in *Cambridge Companion to Aristotle,* Jonathan Barnes, ed. (Cambridge University Press, 1995); Bernard Yack, *The Problems of A Political Animal: Community, Justice, and Conflict in Aristotelian Political Thought* (Berkeley: University of California Press, 1993), 73–74.

43. For a contemporary argument that the particularity of the *polis* relegates Aristotle (and most of ancient Greek political thought) to historical interest alone, see Holmes, "Aristippus." For contrasting arguments, see Yack, *Problems of A Political Animal,* 71–85; J. Peter Euben, *The Tragedy of Political Theory: The Road Not Taken* (Princeton, N.J.: Princeton University Press, 1990), chapter 1.

44. *Politics* 1255b15.

45. *Politics* 1275a20–30, 1275b20. Aristotle is not wholly consistent on this point, however. He does discuss monarchy as a potential type of political rule, thus potentially challenging the understanding of the character of such rule discussed earlier. Nonetheless, I believe that this is the most adequate interpretation

of his overall view. For a helpful discussion of this point, see C.C.W. Taylor, "Politics," 242–247.

46. *Politics* 1280b28.

47. *Politics* 1318b10–15.

48. *Politics* 1252b15–20.

49. *Politics* 1252b28.

50. *Politics* 1252b25–30.

51. *Politics* 1326b3–9. Also Stephen Everson, "Aristotle on the Foundations of the State," *Political Studies* 36 (1988): 94.

52. *Politics* 1253a1.

53. *Politics* 1252b30.

54. *Politics* 1253a20.

55. See Yack, *Problems of a Political Animal,* 93 for a helpful discussion of this. While a defense of this position is surely awkward, it would not necessarily require an identification of the *polis* as "a living organism," as Yack suggests here. Aristotle's discussion of natural body includes nonliving, nonbiological phenomena.

56. *Politics* 1252b30, emphasis added.

57. Leo Strauss, *Natural Right and History* (Chicago: University of Chicago Press, 1953), 7.

58. For instance, quoting from one of Aristotle's early writings, K. von Fritz and E. Kapp maintain that for Aristotle, the lawgiver "derives his measure from 'nature'." See "The Development of Aristotle's Political Philosophy and the Concept of Nature," in Barnes et al., eds., *Articles on Aristotle,* Vol. 2, 116. See also Strauss, *Natural Right and History,* chapters 3 and 4; W. von Leyden, *Aristotle on Equality and Justice* (New York: St. Martin's Press, 1985), 84–90; Paul E. Sigmund, *Natural Law in Political Thought* (Cambridge, Mass.: Winthrop, 1971), 9–12. Fred D. Miller, Jr., "Aristotle on Natural Law and Justice," in Keyt and Miller, eds., *A Companion to Aristotle's* Politics, offers a more nuanced defense of this position.

59. *Politics* 1279a10–20.

60. Miller, "Aristotle on Natural Law and Justice," 300.

61. Fred D. Miller, Jr., "Aristotle's Political Naturalism," *Apeiron* 22 (1989): 217. It should be noted that for Miller, the naturalness of the *polis* is determined by "the extent that it enables its members to realize their natural ends"; thus his argument is distinct from one that attributes an end to the *polis* itself.

62. *Nicomachean Ethics* 1135a5.

63. *Politics* 1325b35.

64. See *Politics* Book 7.

65. Yack, *Problems of a Political Animal,* 90–91.

66. *Nicomachean Ethics* 1134b17–1135a15.

67. For a compelling critique of this treatment of the text, as well as an alternative interpretation, see Bernard Yack, "Natural Right and Aristotle's Understanding of Justice," *Political Theory* 18 (1990): 216–237.

68. See Yack, "Natural Right and Aristotle."

69. Arendt, *Human Condition*, 30.

70. Arendt, *Human Condition*, 30.

71. Arendt, *Human Condition*, 29.

72. Brown, *Manhood and Politics*, 26.

73. For arguments that critically engage Arendt's own perspective here, see' Hanna Fenichel Pitkin, "Justice: Relating Public and Private," *Political Theory* 9 (August 1981): 342–347, and Joseph M. Schwartz, *The Permanence of the Political: A Democratic Critique of the Radical Impulse to Transcend Politics* (Princeton, N.J.: Princeton University Press, 1995), 216, where he argues against Arendt that "although politics may be one of humankind's most creative endeavors, it transpires in a realm of necessity."

74. Arendt, *Human Condition*, 38. Dualism is the subject of analysis by contemporary feminists and environmentalists because it is characterized as a problem to be rooted out. Thus they cannot accept the Arendtian claim that it has disappeared in our modern world without undermining the practical applicability of their project.

75. For an environmentalist interpretation along these lines, see Victor Ferkiss, *Nature, Technology, and Society: Cultural Roots of the Current Environmental Crisis* (New York: New York University Press, 1993), 6; for a feminist one, see Brown, *Manhood and Politics*. For one that seeks to integrate both, see: Plumwood, *Feminism and the Mastery of Nature*. A somewhat surprising and critical adoption of Arendt's characterization of Aristotelian dualism can be found in Murray Bookchin, *The Ecology of Freedom: The Emergence and Dissolution of Hierarchy* (Palo Alto, Calif.: Cheshire Books, 1982), 107. This seems surprising because at many other points Bookchin appeals to Aristotle in a far more positive way as an originator of the "organismic" tradition of philosophy within which he situates himself. Elsewhere, he also celebrates the Athenian *polis* in Arendtian terms in a way that contrasts sharply with this critique of the nature/freedom dualism. Cf. Murray Bookchin, *Remaking Society: Pathways to a Green Future* (Boston: South End Press, 1990), 179–180.

76. Arendt, *Human Condition*, 36–37.

77. *Politics*, 1328a25.

78. See *Politics* 1328a35.

79. *Politics* 1328b15.

80. Arendt, *Human Condition*, 37, emphasis added.

81. See footnote #75 for citations.

82. David Keyt, "Three Basic Theorems in Aristotle's *Politics*," in Keyt and Miller, eds., *A Companion to Aristotle's* Politics, 118.

83. *Politics* 1253a29–31.

84. Barker, *The Politics of Aristotle,* 7n1.

85. Keyt, "Three Basic Theorems," 120.

86. Keyt, "Three Basic Theorems," 120.

87. R.G. Mulgan, *Aristotle's Political Theory* (Oxford: Clarendon Press, 1977), 17.

88. Mulgan, *Aristotle's Political Theory,* 26.

89. *Politics* 1279a30–40.

90. *Politics* 1279b5–10.

91. *Politics* 1275a20; 1277b15.

92. See also *Politics* 1269a35; 1278a5.

93. *Politics* 1328a25.

94. *Politics* 1254b5–20.

95. *Politics* 1255b15.

96. Plumwood, *Feminism and the Mastery of Nature,* 47.

97. Environmentalist writers who stress their affinity with Aristotle on this point include John O'Neill, *Ecology, Policy, and Politics: Human Well-Being and the Natural World* (London: Routledge, 1993), chapter 10; William Ophuls and A. Stephen Boyan, Jr., *Ecology and the Politics of Scarcity Revisited* (New York: W.H. Freeman, 1992), 8; and Mulford Q. Sibley, "The Relevance of Classical Political Theory for Economy, Technology & Ecology," *Alternatives* (Winter 1973): 14–35.

98. See Karl Marx, "On the Jewish Question," in *Marx/Engels Reader,* and Susan Moller Okin, *Justice, Gender, and the Family* (New York: Basic Books, 1989), 124–133, respectively.

99. It is important to note, with Susan Okin, that the claim that the personal is political need not and should not imply the *elimination* of boundaries between public and private, but rather the desirability of negotiating them through public, political deliberations. See Okin, *Justice, Gender and the Family,* x.

100. *Rhetoric* 1360a12–13; *Constitution of Athens* 43.4; also M.I. Finley, *The Ancient Economy,* 2nd ed. (Berkeley: University of California Press, 1985), 162.

101. *Politics* 1276b10.

102. See for example, Martha Nussbaum, "Nature, Function, and Capability: Aristotle on Political Distribution," in *Oxford Studies in Ancient Philosophy,* supplementary volume (Oxford: Clarendon, 1988), 171; and Charles Taylor, "The Nature and Scope of Distributive Justice," in *Justice and Equality, Here and Now,* Frank Lucash, ed. (Ithaca, N.Y.: Cornell University Press, 1986), 37.

103. See footnote 40 in this chapter for a discussion of this point.

104. Martha Nussbaum makes this point, arguing that while Aristotle was empirically wrong in his assessments of the capacities of women and slaves, we can alter this empirical assessment without rejecting his philosophical views. While

Aristotle's misogyny permeates his writings, she argues that neither this attitude nor his position on natural slavery seems to be inherent in his natural philosophy. Nussbaum, "Nature, Function, and Capability," 144–184.

105. *Politics* 1252b30.

106. *Politics* 1254b5.

107. *Politics* Book 7, chapter 4.

Chapter 6

1. For instance, Patrick Riley finds Hobbes to be an inadequate dualist in *Will and Political Legitimacy: A Critical Exposition of Social Contract Theory in Hobbes, Locke, Rousseau, Kant, and Hegel* (Cambridge, Mass.: Harvard University Press, 1982), chapter 2; David Keyt, we have seen, finds Aristotle to be an inadequate derivationist.

2. See John O'Neill, *Ecology, Policy, and Politics: Human Well-being and the Natural World* (London: Routledge, 1993), 23–24, for an argument that also draws upon Aristotle to make this argument.

3. Necessary at least in our present heavily populated, technological world where the constraints of scale alone cannot limit human impact to a sustainable level.

4. Claus Offe, "Challenging the Boundaries of Institutional Politics: Social Movements since the 1960s," in *Changing Boundaries of the Political*, Charles S. Maier, ed. (Cambridge: Cambridge University Press, 1987), 66.

5. For a complementary argument, see Ian Shapiro, *Democratic Justice* (New Haven, Conn.: Yale University Press, 1999), 5–10.

6. See Langdon Winner, *Autonomous Technology: Technics-out-of-control as a Theme in Political Thought* (Cambridge, Mass.: MIT Press, 1977), 237 for a similar point.

7. Andrew Dobson, "Afterword" in *The Politics of Nature: Explorations in Green Political Theory*, Andrew Dobson and Paul Lucardie, eds. (London: Routledge, 1993), 230.

8. *Leviathan*, chapter 30, 229.

9. *Leviathan*, chapter 13, 78.

10. As seen in chapter 2, Bookchin offers a comparable example, which he terms dialectical naturalism, that is more explicitly connected with environmentalist thought.

11. Ronald Beiner, *Political Judgment* (Chicago: University of Chicago Press, 1983), 169n6.

12. For some helpful examples, see Michael Walzer, *The Company of Critics: Social Criticism and Political Commitment in the Twentieth Century* (New York: Basic Books, 1988).

13. *Leviathan*, chapter 13, 78.

14. The term emerged in large part as a consequence of a report by the World Commission on Environment and Development, chaired by former Norwegian Prime Minister Gro Harlem Brundtland, *Our Common Future* (New York: Oxford University Press, 1987).

15. For a discussion of the multiple meanings of sustainable development and the process of change it has undergone, see Sharachchandra M. Lélé, "Sustainable Development: A Critical Review," in *Green Planet Blues: Environmental Politics from Stockholm to Kyoto*, 2nd ed. Ken Conca and Geoffrey D. Dabelko, eds. (Boulder, Col.: Westview Press, 1998) and Andrew Dobson, *Justice and the Environment: Conceptions of Environmental Sustainability and Dimensions of Social Justice* (Oxford: Oxford University Press, 1998), chapter 2.

16. See Larry Lohmann, "Whose Common Future?" in Conca and Dabelko, eds., *Green Planet Blues;* also Frederick H. Buttel, "Environmentalization: Origins, Processes, and Implications for Rural Social Change," *Rural Sociology 57* (1992): 1–27.

17. The best collection of essays in a pragmatic vein, including Weston's, is Andrew Light and Eric Katz, eds., *Environmental Pragmatism* (New York: Routledge, 1996).

18. Anthony Weston, "Beyond Intrinsic Value: Pragmatism in Environmental Ethics," *Environmental Ethics,* 7 (Winter, 1985): 322.

19. Weston, "Beyond Intrinsic Value," 333.

20. Weston, "Beyond Intrinsic Value," 338. Weston completes the argument quoted here by stating "and this is a recognizably factual issue . . . and also negotiable." Here, Weston seems to fail to recognize an important implication of his critique. It is not facts alone that will determine which alternatives are presented as feasible or chosen for adoption. Here, in the definitions of feasibility and criteria for choice, the underlying conceptions of the appropriate character and scope of politics are most influential, although they are rarely acknowledged.

21. Mark Sagoff, *The Economy of the Earth: Philosophy, Law and the Environment* (Cambridge: Cambridge University Press, 1988), 17. While Sagoff advances a Kantian critique of utilitarianism, his pragmatist views (see also pp. 12–14 and pp. 215–216) are most relevant here.

22. Explicated in Sagoff, *Economy of the Earth,* 124–145.

23. Timothy Kaufman-Osborn, *Creatures of Prometheus: Gender and the Politics of Technology* (Lanham, Md.: Rowman and Littlefield, 1997), 145.

24. Mark Sagoff, "Settling America: The Concept of Place in Environmental Politics," in *A Wolf in the Garden: The Land Rights Movement and the New Environmental Debate,* Philip D. Brick and R. McGreggor Cawley, eds. (Lanham, Md.: Rowman and Littlefield, 1996), 252–253. Recent work on place by Bryan Norton is especially relevant and insightful. See "Ecology and Opportunity: Intergenerational Equity and Sustainable Options," in *Fairness and Futurity: Essays on Environmental Sustainability and Social Justice,* Andrew Dobson, ed. (Oxford: Oxford University Press, 1999), 118–150.

25. William Cronon, "The Trouble with Wilderness; or, Getting Back to the Wrong Nature," in *Uncommon Ground: Toward Reinventing Nature* (New York: W.W. Norton, 1995); and Michael P. Nelson, "Rethinking Wilderness: The Need for a New Idea of Wilderness," *Philosophy in the Contemporary World* 3 (Summer 1996): 6–9. For a wide-ranging collection of essays on this theme, see J. Baird Callicott and Michael P. Nelson, eds., *The Great New Wilderness Debate* (Athens: University of Georgia Press, 1998).

26. See James Howard Kunstler, *The Geography of Nowhere: The Rise and Decline of America's Man-Made Landscape* (New York: Simon and Schuster, 1993).

27. See for example, Benjamin R. Barber, *Strong Democracy: Participatory Politics for a New Age* (Berkeley: University of California Press, 1984), 156–158 and passim. For an argument specific to environmental questions, see Michael Saward, "Green Democracy?" in *The Politics of Nature: Explorations in Green Political Theory*, Andrew Dobson and Paul Lucardie, eds. (London: Routledge, 1993), 77.

28. See for example, Riley E. Dunlap, George H. Gallup, Jr., and Alec M. Gallup, "Of Global Concern: Results of the Health of the Planet Survey," *Environment* 35 (November 1993): 7–15, 33–39.

29. I emphasize conceptual marginalization here to distinguish it from the more real (but still far from universal) political marginalization of environmentalism.

Chapter 7

1. *Back To Earth: Tomorrow's Environmentalism* (Philadephia: Temple University Press, 1994), 14.

2. For example, see the essays collected in *Reinventing Nature? Responses to Postmodern Deconstruction,* Michael E. Soulé and Gary Lease, eds. (Washington D.C.: Island Press, 1995), 159.

3. John Muir, *Our National Parks,* Sierra ed., Vol. VI (Boston: Houghton Mifflin, 1917), 3.

4. Warwick Fox, *Toward A Transpersonal Ecology: Developing New Foundations for Environmentalism* (Boston, Shambhala, 1990), 251.

5. See the excellent discussion of this issue in Peter Cannavò, "Timber, Ecology, and Citizenship: Work Practice as a Basis for a Green, Democratic Politics." Paper presented at the Western Political Science Association Meeting, March 27, 1999.

6. See Robert D. Bullard, ed., *Confronting Environmental Racism: Voices from the Grassroots* (Boston: South End Press, 1993); Andrew Szasz, *Ecopopulism: Toxic Waste and the Movement for Environmental Justice* (Minneapolis: University of Minnesota Press, 1994); Daniel Faber, ed., *The Struggle for Ecological Democracy: Environmental Justice Movements in the United States* (New York: Guilford Press, 1998).

7. Quoted in Robert Gottlieb, *Forcing the Spring: The Transformation of the American Environmental Movement* (Washington D.C.: Island Press, 1993), 318.

8. First National People of Color Environmental Leadership Summit, "Principles of Environmental Justice," in *Debating the Earth: The Environmental Politics Reader,* John S. Dryzek and David Schlosberg, eds. (New York: Oxford University Press, 1998), 469.

9. Gottlieb, *Forcing the Spring,* 187.

10. Gottlieb, *Forcing the Spring,* 304.

11. Bullard, *Confronting Environmental Racism,* 30.

12. Gottlieb, *Forcing the Spring,* 210.

13. Gottlieb, *Forcing the Spring,* 209.

14. In this context, Martha Ackelsberg is correct to argue that women's activism often challenges "the very assumption of a dichotomy between public and private, community and workplace." Martha A. Ackelsberg, "Communities, Resistance, and Women's Activism: Some Implications for a Democratic Polity," in *Women and the Politics of Empowerment,* Ann Bookman and Sandra Morgen, eds. (Philadelphia: Temple University Press, 1988), 302.

15. Gottlieb, *Forcing the Spring,* 210.

16. Bullard, *Confronting Environmental Racism,* 206.

17. This is not to suggest that the motivation of "NIMBY" efforts can *never* be impugned. For example, an effort to block subsidized housing that might attract African-American tenants to a white community, or a group home for developmentally disabled residents, might be criticized on the grounds that it is rooted in racism or ignorance. Thus, it is *not* the concern to protect one's home and community that is the subject of criticism, but the basis for perceiving a threat to these that can be appropriately questioned and challenged. Of course, some critics accuse the environmental justice movement of ignorantly believing in environmental risks that do not exist. This argument, too, can be differentiated from a critique of a "NIMBY syndrome" as itself problematic.

18. Gottlieb, *Forcing the Spring,* 234.

19. Mark Dowie, *Losing Ground: American Environmentalism at the Close of the Twentieth Century* (Cambridge, Mass.: MIT Press, 1995), 135.

20. Dowie, *Losing Ground,* 134.

21. Bron Raymond Taylor, "Introduction: The Global Emergence of Popular Ecological Resistance," in *Ecological Resistance Movements,* B.R. Taylor, ed. (Albany: State University of New York Press, 1995), 26.

22. Bron Raymond Taylor, "Popular Ecological Resistance and Radical Environmentalism" in Taylor, ed., *Ecological Resistance Movements,* 334–335.

23. Ramachandra Guha and Juan Martinez-Alier, *Varieties of Environmentalism: Essays North and South* (London: Earthscan, 1997), 23.

24. Guha and Martinez-Alier, *Varieties of Environmentalism,* 17.

25. Chico Mendes, *Fight for the Forest: Chico Mendes in His Own Words.* Additional material by Tony Gross (London: Latin America Bureau, 1989), 41.

26. Coordinating Body for the Indigenous Peoples' Organizations of the Amazon Basin (COICA), "Two Agendas on Amazon Development," in *Green Planet Blues: Environmental Politics from Stockholm to Kyoto,* 2nd ed., Ken Conca and Geoffrey D. Dabelko, eds. (Boulder, Col.: Westview Press, 1998), 341.

27. Guha and Alier, *Varieties of Environmentalism,* 12; Taylor, "Popular Ecological Resistance," 338.

28. Taylor, "Popular Ecological Resistance," 339.

29. For my purposes here I will refer to the "wise use" and "land rights" movements interchangeably, although the former is sometimes taken as a subset of the latter.

30. Philip D. Brick and R. McGreggor Cawley, "Knowing the Wolf, Tending the Garden," in *A Wolf in the Garden: The Land Rights Movement and the New Environmental Debate,* Philip D. Brick and R. McGreggor Cawley, eds. (Lanham, Md.: Rowman and Littlefield, 1996), 7–8.

31. Donald Snow, "The Pristine Silence of Leaving It All Alone," in Brick and Cawley, eds., *A Wolf in the Garden,* 32, 35.

32. Ray Rasker and Jon Roush, "The Economic Role of Environmental Quality in Western Public Lands," in Brick and Cawley, eds., *A Wolf in the Garden,* 191–193.

33. Snow, "Pristine Silence," 36.

34. Snow, "Pristine Silence," 37.

35. Karen Budd-Falen, "Protecting Community Stability and Local Economies," in Brick and Cawley, eds., *A Wolf in the Garden,* 73–76.

36. Snow, "Pristine Silence," 34.

37. Nancie G. Marzulla, "Property Rights Movement: How It Began and Where It Is Headed," in Brick and Cawley, eds., *A Wolf in the Garden,* 39–58. See also John M. Meyer, "Property, Environmentalism, and the Lockean Myth in America: The Challenge of Regulatory Takings," *Proteus: A Journal of Ideas* (15) 2 (Fall 1998): 51–55.

38. Andrew Dobson, *Justice and the Environment: Conceptions of Environmental Sustainability and Dimensions of Social Justice* (Oxford: Oxford University Press, 1998), 25–26.

39. Dobson, *Justice and the Environment,* 24.

40. In the place where I now live, the redwood coast of northern California, the tension and stalemate that can result from this distance is often palpable. Two years of civil disobedience by Julia "Butterfly" Hill, "tree-sitting" 180 feet up in a Pacific Lumber Company–owned redwood tree, crystallized this tension. See the incredibly disparate comments recorded in the local newspaper upon her descent: *http://www.times-standard.com/front/1999/dec/sun_19/callsframe_121999.html*

41. Richard N.L. Andrews, *Managing the Environment, Managing Ourselves: A History of American Environmental Policy* (New Haven, Conn.: Yale University Press, 1999), ix–xii.

Bibliography

Ackelsberg, Martha A. "Communities, Resistance, and Women's Activism: Some Implications for a Democratic Polity," in *Women and the Politics of Empowerment*, Ann Bookman and Sandra Morgen, eds. Philadelphia: Temple University Press, 1988.

Ambler, Wayne H. "Aristotle's Understanding of the Naturalness of the City," *Review of Politics* 47 (April 1985): 163–185.

Anderson, Terry L., and Donald R. Leal. *Free Market Environmentalism*. Boulder, Col.: Westview Press, 1991.

Andrews, Richard N.L. *Managing the Environment, Managing Ourselves: A History of American Environmental Policy*. New Haven, Conn.: Yale University Press, 1999.

Aquinas, Thomas. *Commentary on the Nicomachean Ethics*. Translated by C. I. Litzinger. Chicago: Regnery, 1964.

Arendt, Hannah. *The Human Condition*. Chicago: University of Chicago Press, 1958.

———. *On Revolution*. New York: Viking Press, 1963.

Aristotle. *The Complete Works of Aristotle*. 2 vols. Edited by Jonathan Barnes. Princeton, N.J.: Princeton University Press, 1984.

All discussions, quotations, and footnotes to Aristotle's writings refer to the translations collected in this edition, except for those to the *Politics*, where the Lord translation was primary, and the *Nicomachean Ethics*, where the Ostwald translation was primary. Sachs's *Aristotle's* Physics: *A Guided Study* offers a radically new translation of this work that was consulted extensively. Other translations consulted are listed below.

———. *The Politics of Aristotle*. Translated with an introduction, notes, and appendixes by Ernest Barker. Oxford: Oxford University Press, 1946.

———. *Nicomachean Ethics*. Translated with an introduction and notes by Martin Ostwald. Indianapolis, Ind.: Bobbs-Merrill, 1962.

———. *The Politics*. rev. ed. Translated by T.A. Sinclair, revised and re-presented by Trevor J. Saunders. Harmondsworth, U.K.: Penguin, 1981.

———. *The Politics*. Translated with an introduction, notes, and glossary by Carnes Lord. Chicago: University of Chicago Press. 1984.

Aubrey, John. *Excerpts from Aubrey's Life of Hobbes* from *Brief Lives*. Reprinted in Thomas Hobbes, *Leviathan, with selected variants from the Latin edition of 1668*. Edited with an introduction by Edwin Curley. Indianapolis, Ind.: Hackett, 1994.

Bacon, Francis. *The New Organon, and related writings*. Edited with an introduction by Fulton H. Anderson. New York: Liberal Arts Press, 1960.

Barber, Benjamin R. *Strong Democracy: Participatory Politics for a New Age*. Berkeley: University of California Press, 1984.

———. *The Conquest of Politics*. Princeton, N.J.: Princeton University Press, 1988.

Barbour, Ian, ed. *Western Man and Environmental Ethics: Attitudes toward Nature and Technology*. Reading, Mass.: Addison-Wesley, 1973.

Barnes, Jonathan, ed. *Cambridge Companion to Aristotle*. Cambridge: Cambridge University Press, 1995.

Barnes, Jonathan, Malcolm Schofield, and Richard Sorabji, eds. *Articles on Aristotle*. Vols. 1–4. London: Duckworth, 1975–79.

Barry, John. "The Limits of the Shallow and the Deep: Green Politics, Philosophy, and Praxis." *Environmental Politics* 3 (Autumn 1994): 369–394.

———. *Rethinking Green Politics: Nature, Virtue and Progress*. London: Sage, 1999.

Bartlett, Robert V. "Evaluating Environmental Policy Success and Failure," in *Environmental Policy in the 1990s*, 2nd ed. Norman J. Vig and Michael E. Kraft, eds. Washington D.C.: Congressional Quarterly, 1994.

Baumgold, Deborah. "Hobbes's Political Sensibility: The Menace of Political Ambition," in *Thomas Hobbes and Political Theory*, Mary G. Dietz, ed. Lawrence: University Press of Kansas, 1990.

Becker, Carl. *The Heavenly City of the Eighteenth Century Philosophers*. New Haven, Conn.: Yale University Press, 1959.

Beiner, Ronald. *Political Judgment*. Chicago: University of Chicago Press, 1983.

Bennett, Jane. *Unthinking Faith and Enlightenment: Nature and the State in a Post-Hegelian Era*. New York: New York University Press, 1987.

Bennett, Jane, and William Chaloupka, eds. *In the Nature of Things: Language, Politics, and the Environment*. Minneapolis: University of Minnesota Press, 1993.

Bennett, Jane, and William Chaloupka. "Introduction: TV Dinners and the Organic Brunch," in *In the Nature of Things: Language, Politics and the Environment*, Jane Bennett and William Chaloupka, eds. Minneapolis: University of Minnesota Press, 1993.

Benton, Ted. *Natural Relations: Ecology, Animal Rights and Social Justice*. London: Verso, 1993.

Berry, Wendell. *The Unsettling of America*. San Francisco: Sierra Club Books, 1977.

Bertman, Martin. *Hobbes: The Natural and the Artifacted Good.* Bern: Peter Lang, 1981.

Bobbio, Norberto. *Thomas Hobbes and the Natural Law Tradition.* Translated by Daniella Gobetti. Chicago: University of Chicago Press, 1993.

Bookchin, Murray. "Review of *Ecology as Politics,* by André Gorz," *Telos* 46 (1980–81):176–190.

———. *The Ecology of Freedom: The Emergence and Dissolution of Hierarchy.* Palo Alto, Calif.: Cheshire Books, 1982.

———. *The Rise of Urbanization and the Decline of Citizenship.* San Francisco: Sierra Club Books, 1987.

———. "Social Ecology Versus Deep Ecology," *Socialist Review* 4 (1988): 9–29.

———. *The Philosophy of Social Ecology.* Montréal: Black Rose Books, 1990.

———. "Freedom and Necessity in Nature," in *The Philosophy of Social Ecology: Essays on Dialectical Naturalism,* Murray Bookchin, ed. Montréal: Black Rose Books, 1990.

———. "Introduction: A Philosophical Naturalism," in *The Philosophy of Social Ecology: Essays on Dialectical Naturalism,* Murray Bookchin, ed. Montréal: Black Rose Books, 1990.

———. "Thinking Ecologically," in *The Philosophy of Social Ecology: Essays on Dialectical Naturalism.* Montréal: Black Rose Books, 1990.

———. "Recovering Evolution: A Reply to Eckersley and Fox," *Environmental Ethics* 12 (1990): 253–274.

———. *Remaking Society: Pathways to a Green Future.* Boston: South End Press, 1990.

Booth, William James. "Household and Market: On the Origins of Moral Economic Philosophy," *Review of Politics* 56 (Spring 1994): 207–235.

Botkin, Daniel B. *Discordant Harmonies: A New Ecology for the Twenty-First Century.* New York: Oxford University Press, 1990.

Bowle, John. *Hobbes and His Critics: A Study in Seventeenth Century Constitutionalism.* Oxford: Jonathan Cape, 1951.

Bradford, George. *How Deep is Deep Ecology?* Ojai, Calif.: Times Change Press, 1989.

Bramwell, Anna. *Ecology in the 20th Century: A History.* New Haven, Conn.: Yale University Press, 1989.

Brandt, Frithiof. *Thomas Hobbes' Mechanical Conception of Nature.* Copenhagen and London: Levin & Munksgaard and Hachette, 1928.

Brennan, Andrew. *Thinking about Nature.* Athens: University of Georgia Press, 1988.

Brick, Philip D., and R. McGreggor Cawley. "Knowing the Wolf, Tending the Garden," in *A Wolf in the Garden: The Land Rights Movement and the New Environmental Debate,* Philip D. Brick and R. McGreggor Cawley, eds. Lanham, Md.: Rowman and Littlefield, 1996.

Brown, Keith C., ed. *Hobbes Studies*. Oxford: Blackwell, 1965.

Brown, Lester R., and others. *State of the World*. New York: W.W. Norton, annual report.

Brown, Wendy L. *Manhood and Politics: A Feminist Reading in Political Theory*. Totowa, N.J.: Rowman and Littlefield, 1988.

Brundtland, Gro Harlem. *Our Common Future*. New York: Oxford University Press, 1987.

Budd-Falen, Karen. "Protecting Community Stability and Local Economies," in *A Wolf in the Garden: The Land Rights Movement and the New Environmental Debate*, Philip D. Brick and R. McGreggor Cawley, eds. Lanham, Md.: Rowman and Littlefield, 1996.

Bullard, Robert D., ed. *Confronting Environmental Racism: Voices from the Grassroots*. Boston: South End Press, 1993.

Burtt, Edwin Arthur. *The Metaphysical Foundations of Modern Physical Science*. rev. ed. London: Routledge & Kegan Paul, 1932.

Buttel, Frederick H. "Environmentalization: Origins, Processes, and Implications for Rural Social Change," *Rural Sociology* 57 (1992): 1–27.

Callicott, J. Baird. "Animal Liberation: A Triangular Affair," *Environmental Ethics* 2 (1980): 311–338.

———, ed. *Companion to* A Sand County Almanac. Madison: University of Wisconsin Press, 1987.

———. "The Land Aesthetic," in *Companion to* A Sand County Almanac. Madison: University of Wisconsin Press, 1987.

———. *Earth's Insights*. Berkeley: University of California Press, 1994.

———. "Hume's Is/Ought Dichotomy and the Relation of Ecology to Leopold's Land Ethic," *Environmental Ethics* 4 (Summer 1982): 163–174.

———. "Intrinsic Value, Quantum Theory and Environmental Ethics," *Environmental Ethics* 7 (1985): 257–275.

———. "The Case against Moral Pluralism," *Environmental Ethics* 12 (1990): 99–124.

Callicott, J. Baird, and Michael P. Nelson, eds. *The Great New Wilderness Debate*. Athens: University of Georgia Press, 1998.

Cannavò, Peter. "Timber, Ecology, and Citizenship: Work Practice as a Basis for a Green, Democratic Politics." Paper presented at the Western Political Science Association Meeting, March 27, 1999.

Capra, Fritjof. *The Turning Point: Science, Society, and the Rising Culture*. New York: Bantam Books, 1982.

Carter, Paul. *The Road to Botany Bay: An Exploration of Landscape and History*. Chicago: University of Chicago Press, 1987.

Catton, William R., Jr. *Overshoot: The Ecological Basis of Revolutionary Change*. Urbana: University of Illinois Press, 1982.

Clarke, J.J. *Voices of the Earth: An Anthology of Ideas and Arguments.* New York: George Braziller, 1994.

Coates, Peter. *Nature: Western Attitudes since Ancient Times.* Berkeley: University of California Press, 1998.

Cohen, I. Bernard. *Interactions: Some Contacts between the Natural Sciences and the Social Sciences.* Cambridge, Mass.: MIT Press, 1994.

Collingwood, R.G. *The Idea of Nature.* Oxford: Oxford University Press, 1945.

Commoner, Barry. *The Closing Circle: Nature, Man and Technology.* New York: Bantam, 1971.

———. "Why We Have Failed," *Greenpeace* (Sept./Oct. 1989): 12–13.

Coordinating Body for the Indigenous Peoples' Organizations of the Amazon Basin (COICA). "Two Agendas on Amazon Development," in *Green Planet Blues: Environmental Politics from Stockholm to Kyoto,* 2nd ed. Ken Conca and Geoffrey D. Dabelko, eds.. Boulder, Col.: Westview Press, 1998.

Cox, Harvey. *The Secular City.* New York: Macmillan, 1965.

Cranston, Maurice, and Richard Peters, eds. *Hobbes and Rousseau: A Collection of Critical Essays.* Garden City, N.Y.: Anchor Books, 1972.

Cronon, William. *Nature's Metropolis: Chicago and the Great West.* New York: W.W. Norton, 1991.

———, ed. *Uncommon Ground: Toward Reinventing Nature.* New York: W.W. Norton, 1995.

———. "The Trouble with Wilderness; or, Getting Back to the Wrong Nature," in *Uncommon Ground: Toward Reinventing Nature,* William Cronon, ed. New York: W.W. Norton, 1995.

Daly, Herman E., and John B. Cobb, Jr. *For The Common Good: Redirecting the Economy Toward Community, the Environment, and a Sustainable Future.* Boston: Beacon Press, 1989.

Descartes, René. *Meditations on First Philosophy: with Selections from the Objections and Replies.* Translated by John Cottingham. Cambridge: Cambridge University Press, 1986.

de-Shalit, Avner. "From the Political to the Objective: The Dialectics of Zionism and The Environment," *Environmental Politics* 4 (Spring 1995): 70–87.

———. "Is Liberalism Environment-Friendly?" *Social Theory and Practice* 21 (Summer 1995): 287–314.

Devall, Bill. "Earthday 25: A Retrospective of Reform Environmental Movements," *Philosophy in the Contemporary World* 2 (Winter 1995): 9–15.

Devall, Bill, and George Sessions. *Deep Ecology: Living As If Nature Mattered.* Salt Lake City: Peregrine Smith, 1985.

Dewey, John. *Reconstruction in Philosophy.* New York: Henry Holt, 1920.

Dickens, Peter. *Society and Nature: Toward a Green Social Theory.* Philadelphia: Temple University Press, 1992.

Dietz, Mary. "Hobbes's Subject as Citizen," in *Thomas Hobbes and Political Theory*, Mary Dietz, ed. Lawrence: University Press of Kansas, 1990.

diZerega, Gus. "Empathy, Society, Nature and the Relational Self: Deep Ecology and Liberal Modernity," *Social Theory and Practice* 21 (Summer 1995): 239–269.

———. "Social Ecology, Deep Ecology, and Liberalism," *Critical Review* 6 (Spring-Summer 1992): 305–370.

Dobson, Andrew. *Green Political Thought*. London: Unwin Hyman, 1990.

Dobson, Andrew, and Paul Lucardie, eds. *The Politics of Nature: Explorations in Green Political Theory*. London: Routledge, 1993.

———. "Afterword." in *The Politics of Nature: Explorations in Green Political Theory*, Andrew Dobson and Paul Lucardie, eds. London: Routledge, 1993.

———. *Green Political Thought*. 2nd ed. London: Routledge, 1995.

———. *Justice and the Environment: Conceptions of Environmental Sustainability and Dimensions of Social Justice*. Oxford: Oxford University Press, 1998.

———, ed. *Fairness and Futurity: Essays on Environmental Sustainability and Social Justice*. Oxford: Oxford University Press, 1999.

Doherty, Brian, and Marius de Geus, eds. *Democracy and Green Political Thought: Sustainability, Rights and Citizenship*. London: Routledge, 1996.

Dowie, Mark. *Losing Ground: American Environmentalism at the Close of the Twentieth Century*. Cambridge, Mass.: MIT Press, 1995.

Drengson, Alan R. *Beyond Environmental Crisis: From Technocrat to Planetary Person*. New York: Peter Lang, 1989.

Dryzek, John. *Rational Ecology: Environment and Political Economy*. New York: Basil Blackwell, 1987.

Dunlap, Riley E. "Trends in Public Opinion toward Environmental Issues: 1965–1990," in *American Environmentalism: The U.S. Environmental Movement, 1970–1990*, Riley E. Dunlap and Angela G. Mertig, eds. Philadelphia: Taylor and Francis, 1990.

Dunlap, Riley E., George H. Gallup, Jr., and Alec M. Gallup. "Of Global Concern: Results of the Health of the Planet Survey," *Environment* 35 (November 1993): 7–15, 33–39.

Easterbrook, Gregg. *A Moment on the Earth: The Coming Age of Environmental Optimism*. New York: Viking, 1995.

Eckersley, Robin. "Green Politics: A Practice in Search of a Theory?" *Alternatives* 15 (1988): 52–61.

———. "Divining Evolution: The Ecological Ethics of Murray Bookchin." *Environmental Ethics* 11 (1989): 99–116.

———. *Environmentalism and Political Theory: Toward an Ecocentric Approach*. Albany: State University of New York Press, 1992.

Ehrenfeld, David. *The Arrogance of Humanism*. Oxford: Oxford University Press, 1981.

Enzensberger, Hans-Magnus. "A Critique of Political Ecology," *New Left Review* 84 (March-April 1974): 3–31.

Euben, J. Peter. *The Tragedy of Political Theory: The Road Not Taken.* Princeton, N.J.: Princeton University Press, 1990.

Evernden, Neil. *The Social Creation of Nature.* Baltimore, Md.: Johns Hopkins University Press, 1992.

Everson, Stephen. "Aristotle on the Foundations of the State," *Political Studies* 36 (1988): 89–101.

Faber, Daniel, ed. *The Struggle for Ecological Democracy: Environmental Justice Movements in the United States.* New York: Guilford Press, 1998.

Ferkiss, Victor. *Nature, Technology, Society: Cultural Roots of the Current Environmental Crisis.* New York: New York University Press, 1993.

Ferry, Luc. *The New Ecological Order.* Translated by Carol Volk. Chicago: University of Chicago Press, 1995.

Finley, M. I. "Aristotle and Economic Analysis." Chap. in *Studies in Ancient Society.* London: Routledge and Kegan Paul, 1974.

———. *Politics in the Ancient World.* Cambridge: Cambridge University Press, 1983.

———. *The Ancient Economy.* 2d ed. Berkeley: University of California Press, 1985.

First National People of Color Environmental Leadership Summit. "Principles of Environmental Justice," in *Debating the Earth: The Environmental Politics Reader,* John S. Dryzek and David Schlosberg, eds. New York: Oxford University Press, 1998.

Fischer, Frank, and Michael Black, eds. *Greening Environmental Policy: The Politics of a Sustainable Future.* New York: St. Martins, 1995.

Fox, Warwick. "The Deep Ecology-Ecofeminism Debate and its Parallels," *Environmental Ethics* 11 (Spring 1989): 5–25.

———. *Toward a Transpersonal Ecology: Developing New Foundations for Environmentalism.* Boston: Shambhala, 1990.

Fritzell, Peter A. "The Conflicts of an Ecological Conscience," in *Companion to A Sand County Almanac,* J. Baird Callicott, ed. Madison: University of Wisconsin Press, 1987.

Galileo. *The Assayer,* in *Discoveries and Opinions of Galileo.* Translation with introduction and notes by Stillman Drake. New York: Anchor, 1957.

Garrett, Jan. "Aristotle, Ecology and Politics: Theoria and Praxis for the Twenty-first Century," in *Communitarianism, Liberalism, and Social Responsibility,* Creighton Peden and Yeager Hudson, eds. Lewiston, N.Y.: Edwin Mellen Press, 1992.

Glacken, Clarence J. *Traces on the Rhodian Shore: Nature and Culture in Western Thought from Ancient Times to the End of the Eighteenth Century.* Berkeley: University of California Press, 1967.

Goldsmith, M. M. *Hobbes's Science of Politics*. New York: Columbia University Press, 1966.

Goodin, Robert. *Green Political Theory*. Cambridge, U.K.: Polity Press, 1992.

——. "Enfranchising the Earth, and its Alternatives," *Political Studies* 44 (1996): 835–849.

Gorz, André. *Ecology as Politics*. Boston: South End Press, 1980.

Gottlieb, Robert. *Forcing the Spring: The Transformation of the American Environmental Movement*. Washington D.C.: Island Press, 1993.

Gray, John. "An Agenda for Green Conservatism," chap. in *Beyond the New Right: Markets, Government and the Common Environment*. New York: Routledge, 1993.

Guha, Ramachandra, and Juan Martinez-Alier. *Varieties of Environmentalism: Essays North and South*. London: Earthscan, 1997.

Hardin, Garrett. "The Tragedy of the Commons," *Science* 162 (1968): 1243–1248.

Hargrove, Eugene C. *Foundations of Environmental Ethics*. Englewood Cliffs, N.J.: Prentice Hall, 1989.

Hayward, Tim. *Ecological Thought: An Introduction*. Cambridge, U.K.: Polity Press, 1994.

Heilbroner, Robert. *An Inquiry into the Human Prospect: Updated and Reconsidered for the 1980s*. New York: W.W. Norton, 1980.

Hinchman, Lewis P., and Sandra K. "'Deep Ecology' and the Revival of Natural Right," *Western Political Quarterly* 42 (Fall 1989): 201–228.

Hobbes, Thomas. *The English Works of Thomas Hobbes*. Edited by Sir William Molesworth. Vol.1, *Elements of Philosophy, the first section, concerning Body*. London: 1839.

——. *The English Works of Thomas Hobbes*. Edited by Sir William Molesworth. Vol.5, *The Questions Concerning Liberty, Necessity and Chance*. London: 1839.

——. *Behemoth: or, The Long Parliament*. Edited by Ferdinand Tönnies, with an introduction by Stephen Holmes. Chicago: University of Chicago Press, 1990.

——. *De Cive, also known as Philosophical Rudiments Concerning Government and Society*, in *Man and Citizen*. Edited with an introduction by Bernard Gert. Indianapolis, Ind.: Hackett, 1991.

——. *De Homine*. Portions in *Man and Citizen*. Edited with an introduction by Bernard Gert. Indianapolis, Ind.: Hackett, 1991.

——. *Human Nature and De Corpore Politico*. Edited with an introduction by J.C.A. Gaskin. Oxford: Oxford University Press, 1994.

——. *Leviathan, with selected variants from the Latin edition of 1668*. Edited with an introduction by Edwin Curley. Indianapolis, Ind.: Hackett, 1994.

——. *Three Discourses: A Critical Modern Edition of Newly Identified Work of the Young Hobbes*. Edited by Noel B. Reynolds and Arlene W. Saxonhouse. Chicago: University of Chicago Press, 1995.

[Hobbes, Thomas and] René Descartes. "The Third Set of Objections with the Author's Reply," in *Philosophical Works of Descartes*. Edited and translated by Elizabeth S. Haldane and G.R.T. Ross. Cambridge: Cambridge University Press, 1911.

Hoffert, Robert W. "The Scarcity of Politics: Ophuls and Western Political Thought," *Environmental Ethics* 8 (1986): 5–32.

Holmes, Steven Taylor. "Aristippus in and out of Athens," *American Political Science Review* 73 (1979): 113–138.

Honig, Bonnie. *Political Theory and the Displacement of Politics*. Ithaca, N.Y.: Cornell University Press, 1993.

Hughes, J. Donald. "Ecology in Ancient Greece," *Inquiry* 18 (1975): 115–125.

Kargon, Robert Hugh. *Atomism in England from Hariot to Newton*. Oxford: Clarendon Press, 1966.

Kaufman-Osborn, Timothy. *Creatures of Prometheus: Gender and the Politics of Technology*. Lanham, Md.: Rowman and Littlefield, 1997.

Keyt, David. "Three Basic Theorems in Aristotle's *Politics*," in *A Companion to Aristotle's* Politics. Cambridge, Mass.: Blackwell, 1991.

Keyt, David, and Fred D. Miller, Jr., eds. *A Companion to Aristotle's* Politics. Cambridge, Mass.: Blackwell, 1991.

————, and ————. "Introduction," in *A Companion to Aristotle's* Politics, David Keyt and Fred D. Miller, Jr., eds. Cambridge, Mass.: Blackwell, 1991.

Koyré, Alexandre. *From the Closed World to the Infinite Universe*. Baltimore, Md.: Johns Hopkins University Press, 1957.

————. *Metaphysics and Measurement: Essays in Scientific Revolution*. Cambridge, Mass.: Harvard University Press, 1968.

————. "Galileo and Plato," in *Metaphysics and Measurement: Essays in Scientific Revolution*, Alexandre Koyré, ed. Cambridge, Mass.: Harvard University Press, 1968.

————. "The Significance of the Newtonian Synthesis," chap. in *Newtonian Studies*. Cambridge, Mass.: Harvard University Press, 1965.

Kraft, Michael E., and Norman J. Vig. "Environmental Policy from the Seventies to the Nineties: Continuity and Change," in *Environmental Policy in the 1990s*, Michael E. Kraft and Norman J. Vig, eds. Washington, D.C.: Congressional Quarterly Press, 1990.

Kuhn, Thomas S. *The Structure of Scientific Revolutions*. 2nd ed. Chicago: University of Chicago Press, 1970.

Kunstler, James Howard. *The Geography of Nowhere: The Rise and Decline of America's Man-Made Landscape*. New York: Simon and Schuster, 1993.

Lappé, Frances Moore, and J. Baird Callicott, "Marx Meets Muir: Toward a Synthesis of the Progressive Political and Ecological Visions," *Tikkun* 2(4): 16–21.

Latour, Bruno. *We Have Never Been Modern*. Translated by Catherine Porter. Cambridge, Mass.: Harvard University Press, 1993.

Leeson, Susan M. "Philosophic Implications of the Ecological Crisis: The Authoritarian Challenge to Liberalism," *Polity* 11 (1979): 303–318.

Leopold, Aldo. *A Sand County Almanac, with Essays on Conservation from Round River.* New York: Ballantine Books, 1970.

———. "[1947] Forward," in *Companion to A Sand County Almanac,* J. Baird Callicott, ed. Madison: University of Wisconsin Press, 1987.

Lewis, C.S. *Studies in Words.* Cambridge: Cambridge University Press, 1960.

Light, Andrew, and Eric Katz, eds. *Environmental Pragmatism.* New York: Routledge, 1996.

Lindberg, David C. *The Beginnings of Western Science: The European Scientific Tradition in Philosophical, Religious, and Institutional Context, 600 B.C. to A.D. 1450.* Chicago: University of Chicago Press, 1992.

Lipschutz, Ronnie D., with Judith Mayer. *Global Civil Society and Global Environmental Governance: The Politics of Nature from Place to Planet.* Albany: State University of New York Press, 1996.

Locke, John. *Two Treatises of Government.* Edited by Peter Laslett. Cambridge: Cambridge University Press, 1988.

Lohmann, Larry. "Whose Common Future?" in *Green Planet Blues: Environmental Politics from Stockholm to Rio,* Ken Conca, Michael Alberty, and Geoffrey D. Dabelko, eds. Boulder, Col.: Westview Press, 1995.

Lovejoy, Arthur O. *The Great Chain of Being: A Study of the History of an Idea.* Cambridge, Mass.: Harvard University Press, 1936.

Lovelock, James. *The Ages of Gaia: A Biography of Our Living Earth.* New York: Bantam, 1990.

Lynch, Tony, and David Wells. "Environmentalism and the Social Contract," *Political Theory Newsletter* (Australia) 8 (1996): 1–10.

Macauley, David, ed. *Minding Nature: The Philosophers of Ecology.* New York: Guilford Press, 1996.

Machan, Tibor R. "Pollution and Political Theory," in *Earthbound: New Introductory Essays in Environmental Ethics,* Tom Regan, ed. New York: Random House, 1984.

MacIntyre, Alasdair. *After Virtue: A Study in Moral Theory.* 2nd ed. Notre Dame, Ind.: University of Notre Dame Press, 1984.

MacPherson, C.B. *The Political Theory of Possessive Individualism.* Oxford: Oxford University Press, 1962.

Maier, Charles S. *Changing Boundaries of the Political: Essays on the Evolving Balance between the State and Society, Public and Private in Europe.* New York: Cambridge University Press, 1987.

Malcolm, Noel. "Hobbes and the Royal Society," in *Perspectives on Thomas Hobbes,* G.A.J. Rogers and Alan Ryan, eds. Oxford: Clarendon Press, 1988.

Marshall, Peter. *Nature's Web: An Exploration of Ecological Thinking.* London: Simon and Schuster, 1992.

Martell, Luke. *Ecology and Society: An Introduction.* Amherst: University of Massachusetts Press, 1994.

Marx, Karl. "On the Jewish Question," in *The Marx-Engels Reader.* Edited by Robert Tucker. New York: W.W. Norton, 1978.

Marx, Karl, and Frederich Engels. *The Marx-Engels Reader.* Edited by Robert Tucker. New York: W.W. Norton, 1978.

Marx, Leo. *The Machine in the Garden: Technology and the Pastoral Ideal in America.* New York: Oxford University Press, 1964.

Marzulla, Nancie G. "Property Rights Movement: How It Began and Where It Is Headed," in *A Wolf in the Garden: The Land Rights Movement and the New Environmental Debate,* Philip D. Brick and R. McGreggor Cawley, eds. Lanham, Md.: Rowman and Littlefield, 1996.

Mathews, Freya. *The Ecological Self.* Savage, Md.: Barnes and Noble, 1991.

———, ed. *Ecology and Democracy.* London: Frank Cass, 1996.

McCloskey, H.J. *Ecological Ethics and Politics.* Totowa, N.J.: Rowman and Littlefield, 1983.

McLaughlin, Andrew. "Images and Ethics of Nature," *Environmental Ethics* 7 (Winter 1985): 293–319.

———. *Regarding Nature: Industrialism and Deep Ecology.* Albany: State University of New York Press, 1993.

Meadows, Donella H., Dennis M. Meadows, and Jorgen Ranters III, *The Limits to Growth.* New York: Universe Books, 1972.

Meier, Christian. *The Greek Discovery of Politics.* Translated by David McLintock. Cambridge, Mass.: Harvard University Press, 1990.

Mendes, Chico. *Fight for the Forest: Chico Mendes in His Own Words.* Additional material by Tony Gross. London: Latin America Bureau, 1989.

Merchant, Carolyn. *The Death of Nature: Women, Ecology and the Scientific Revolution.* New York: HarperCollins, 1989.

———. *Radical Ecology: The Search for a Livable World.* New York: Routledge, 1992.

Metz, Johannes. *Theology of the World.* New York: Herder and Herder, 1969.

Meyer, John M. "Gifford Pinchot, John Muir, and the Boundaries of Politics in American Thought," *Polity* 30 (Winter 1997): 267–284.

———. "Property, Environmentalism, and the Lockean Myth in America: The Challenge of Regulatory Takings," *Proteus: A Journal of Ideas* 15 (Fall 1998): 51–55.

———. "Interpreting Nature and Politics in the History of Western Thought: The Environmentalist Challenge," *Environmental Politics* 8 (Summer 1999): 1–23.

———. "Rights to Life?: On Nature, Property and Biotechnology," *Journal of Political Philosophy* 8 (June 2000): 154–176.

———. "What is Environmental Political Theory?" *Political Theory,* 29 (April 2001): 276–288.

Milbrath, Lester W. "The World is Relearning Its Story about How the World Works," in *Environmental Politics in the International Arena: Movements, Parties, Organizations and Policy,* Sheldon Kamieniecki, ed. Albany: State University of New York Press, 1993.

Mill, John Stuart. "Nature," in *Nature and the Utility of Religion.* Indianapolis, Ind.: Bobbs-Merrill, 1958.

Miller, Fred D., Jr. "Aristotle's Political Naturalism," *Apeiron* 22 (1989): 195–218.

———. "Aristotle on Natural Law and Justice," in *A Companion to Aristotle's Politics.* Cambridge, Mass.: Blackwell, 1991.

Mintz, Samuel I. *The Hunting of Leviathan: Seventeenth Century Reaction to the Materialism and Moral Philosophy of Thomas Hobbes.* Cambridge: Cambridge University Press, 1962.

Muir, John. *Our National Parks.* Sierra ed., Vol. VI. Boston: Houghton Mifflin, 1917.

Mulgan, R.G. *Aristotle's Political Theory.* Oxford: Clarendon Press, 1977.

Naess, Arne. "The Shallow and the Deep, Long-Range Ecology Movement: A Summary," *Inquiry* 16 (1973): 95–100.

———. *Ecology, Community, Lifestyle: Outline of An Ecosophy.* Translated and revised by David Rothenberg. Cambridge: Cambridge University Press, 1989.

Nash, Roderick. "Aldo Leopold's Intellectual Heritage," in *Companion to A Sand County Almanac,* J. Baird Callicott, ed. Madison: University of Wisconsin Press, 1987.

———. *The Rights of Nature: A History of Environmental Ethics.* Madison: University of Wisconsin Press, 1989.

Nederman, Cary J. "The Puzzle of the Political Animal: Nature and Artifice in Aristotle's Political Theory," *Review of Politics* 56 (Spring 1994): 283–304.

Nelson, Michael P. "Rethinking Wilderness: The Need for a New Idea of Wilderness," *Philosophy in the Contemporary World* 3 (Summer 1996): 6–9.

Norton, Bryan G. *Toward Unity among Environmentalists.* New York: Oxford University Press, 1991.

———. "Ecology and Opportunity: Intergenerational Equity and Sustainable Options," in *Fairness and Futurity: Essays on Environmental Sustainability and Social Justice,* Andrew Dobson, ed. Oxford: Oxford University Press, 1999, pp. 118–150.

Nussbaum, Martha. "Nature, Function, and Capability: Aristotle on Political Distribution," in *Oxford Studies in Ancient Philosophy,* supplementary volume, Oxford: Clarendon Press, 1988, pp. 144–184.

Oakeshott, Michael. "Introduction to *Leviathan*," chap. in *Rationalism in Politics and Other Essays.* New and expanded ed. Indianapolis, Ind.: Liberty Press, 1991.

————. "Logos and Telos," chap. in *Rationalism in Politics and Other Essays*. New and expanded ed. Indianapolis, Ind.: Liberty Press, 1991.

————. "The Moral Life in the Writings of Thomas Hobbes," chap. in *Rationalism in Politics and Other Essays*. New and expanded ed. Indianapolis, Ind.: Liberty Press, 1991.

O'Connor, James J. "Capitalism, Nature, Socialism: A Theoretical Introduction," *Capitalism, Nature, Socialism* 1 (Fall 1988): 11–38.

Offe, Claus. "Challenging the Boundaries of Institutional Politics: Social Movements since the 1960s," in *Changing Boundaries of the Political*, Charles S. Maier, ed. New York: Cambridge University Press, 1987.

Okin, Susan Moller. *Women in Western Political Thought*. Princeton, N.J.: Princeton University Press, 1979.

————. *Justice, Gender and the Family*. New York: Basic Books, 1989.

O'Neill, John. *Ecology, Policy and Politics: Human Well-Being and the Natural World*. New York: Routledge, 1993.

Ophuls, William. "Leviathan or Oblivion?" in *Toward a Steady State Economy*, Herman E. Daly, ed. San Francisco: W.H. Freeman, 1973.

————. *Ecology and the Politics of Scarcity: Prologue to a Political Theory of the Steady State*. San Francisco: W.H. Freeman, 1977.

————. *Requiem for Modern Politics: The Tragedy of the Enlightenment and the Challenge of the New Millennium*. Boulder, Col.: Westview Press, 1997.

Ophuls, William, and A. Stephen Boyan, Jr. *Ecology and the Politics of Scarcity Revisited: The Unraveling of the American Dream*. New York: W.H. Freeman, 1992.

Orr, David W., and Stuart Hill. "Leviathan, The Open Society, and the Crisis of Ecology," *Western Political Quarterly* 31 (1978): 457–469.

Paehlke, Robert C. *Environmentalism and the Future of Progressive Politics*. New Haven, Conn.: Yale University Press, 1989.

Passmore, John. *Man's Responsibility for Nature: Ecological Problems and Western Traditions*. New York: Scribner's, 1974.

Peluso, Nancy Lee. "Coercing Conservation: The Politics of State Resource Control," in *The State and Social Power in Global Environmental Politics*, Ronnie D. Lipschutz and Ken Conca, eds. New York: Columbia University Press, 1993.

Pepper, David. "Anthropocentrism, Humanism and Eco-Socialism: A Blueprint for the Survival of Ecological Politics," *Environmental Politics* 2 (Autumn 1993): 428–452.

————. *Eco-Socialism: From Deep Ecology to Social Justice*. New York: Routledge, 1993.

Peters, Richard. *Hobbes*. Harmondsworth, U.K.: Penguin, 1956.

Pickering, Kevin T., and Lewis A. Owen. *An Introduction to Global Environmental Issues*. London: Routledge, 2nd ed., 1997.

Pitkin, Hanna Fenichel. "Justice: On Relating Public and Private," *Political Theory* 9 (August 1981): 327–352.

Plato. *The Republic.* Translated by Allan Bloom. New York: Basic Books, 1968.

Plumwood, Val. "Ecofeminism: An Overview and Discussion of Positions and Arguments," *Australasian Journal of Philosophy* Supplement to 64 (June 1986): 120–138.

————. *Feminism and the Mastery of Nature.* London: Routledge, 1993.

————. "Inequality, Ecojustice, and Ecological Rationality," reprinted in *Debating the Earth: The Environmental Politics Reader,* ed. John S. Dryzek and David Schlosberg, eds. Oxford: Oxford University Press, 1998, pp. 559–583.

Pois, Robert A. *National Socialism and the Religion of Nature.* London: Croon Helm, 1986.

Rapaczynski, Andrzej. *Nature and Politics: Liberalism in the Philosophies of Hobbes, Locke and Rousseau.* Ithaca, N.Y.: Cornell University Press, 1986.

Rasker, Ray, and Jon Roush. "The Economic Role of Environmental Quality in Western Public Lands," in *A Wolf in the Garden: The Land Rights Movement and the New Environmental Debate,* Philip D. Brick and R. McGreggor Cawley, eds. Lanham, Md.: Rowman and Littlefield, 1996.

Riley, Patrick. *Will and Political Legitimacy: A Critical Exposition of Social Contract Theory in Hobbes, Locke, Rousseau, Kant, and Hegel.* Cambridge, Mass.: Harvard University Press, 1982.

Robertson, George Croom. *Hobbes.* Edinburgh: Blackwood, 1886.

Rodman, John. "Paradigm Change in Political Science: An Ecological Perspective," *American Behavioral Scientist* 24 (1980): 49–74.

————. "Four Forms of Ecological Consciousness Reconsidered," in *Ethics and the Environment,* Donald Scherer and Thomas Attig, eds. Englewood Cliffs, N.J.: Prentice Hall, 1983.

Rogers, G.A.J., and Alan Ryan, eds. *Perspectives on Thomas Hobbes.* Oxford: Clarendon Press, 1988.

Rolston, Holmes III. "Can and Ought We to Follow Nature?" *Environmental Ethics* 1 (Spring 1979): 7–30.

Ross, Andrew. *The Chicago Gangster Theory of Life: Nature's Debt to Society.* London: Verso, 1994.

Ross, W. D. *Aristotle.* 5th ed. London: Methuen, 1949.

Rousseau, Jean-Jacques. *Discourse on the Origin and Foundations of Inequality,* in *The First and Second Discourses.* Edited by Roger D. Masters and translated by Roger D. and Judith R. Masters. New York: St. Martin's Press, 1964.

Routley, Richard (see also Sylvan, Richard). "Roles and Limits of Paradigms in Environmental Thought and Action," in *Environmental Philosophy,* Robert Elliot and Arran Gare, eds. New York: University of Queensland Press, 1983.

Rubin, Charles T. *The Green Crusade: Rethinking the Roots of Environmentalism.* New York: Free Press, 1994.

Ryle, Martin. *Ecology and Socialism*. London: Radius, 1988.

Sachs, Joe. *Aristotle's* Physics: *A Guided Study*. New Brunswick, N.J.: Rutgers University Press, 1995.

Sagoff, Mark. *The Economy of the Earth: Philosophy, Law and the Environment*. Cambridge: Cambridge University Press, 1988.

———. "Settling America: The Concept of Place in Environmental Politics," in *A Wolf in the Garden: The Land Rights Movement and the New Environmental Debate*, Philip D. Brick and R. McGreggor Cawley, eds. Lanham, Md.: Rowman and Littlefield, 1996.

Sale, Kirkpatrick. *Dwellers in the Land: The Bioregional Vision*. San Francisco: Sierra Club Books, 1985.

Salleh, Ariel Kay. "Deeper than Deep Ecology: The Eco-Feminist Connection," *Environmental Ethics* 6 (Winter 1984): 339–345.

Salleh, Ariel. *Ecofeminism as Politics: Nature, Marx and the Postmodern*. London: Zed Books, 1997.

Sandel, Michael J. "The Procedural Republic and the Unencumbered Self," *Political Theory* 12 (February 1984): 81–96.

Sandilands, Catriona. "From Natural Identity to Radical Democracy," *Environmental Ethics* 17 (Spring 1995): 74–91.

Saward, Michael. "Green Democracy?" in *The Politics of Nature: Explorations in Green Political Theory*, Andrew Dobson and Paul Lucardie, eds. London: Routledge, 1993.

Saxonhouse, Arlene. "Aristotle: Defective Males, Hierarchy and the Limits of Politics," in *Women in the History of Political Thought: Ancient Greece to Machiavelli*. New York: Praeger, 1985.

Schumacher, E.F. *Small Is Beautiful: Economics as if People Mattered*. New York: Harper and Row, 1973.

Schwartz, Joseph M. *The Permanence of the Political: A Democratic Critique of the Radical Impulse to Transcend Politics*. Princeton, N.J.: Princeton University Press, 1995.

Serres, Michel. *The Natural Contract*. Translated by Elizabeth MacArthur and William Paulson. Ann Arbor: University of Michigan Press, 1995.

Sessions, George S. "Anthropocentrism and the Environmental Crisis," *Humboldt Journal of Social Relations* 2 (1974): 71–81.

———. "Ecological Consciousness and Paradigm Change," in *Deep Ecology*, Michael Tobias, ed. San Diego: Avant Books, 1985.

Shapin, Steven, and Simon Schaffer. *Leviathan and the Air-Pump: Hobbes, Boyle, and the Experimental Life*. Princeton, N.J.: Princeton University Press, 1985.

Shapiro, Ian. *Democratic Justice*. New Haven, Conn.: Yale University Press, 1999.

Sheldrake, Rupert. *The Rebirth of Nature: The Greening of Science and God*. London: Century, 1990.

Shiva, Vandana. *Staying Alive: Women, Ecology and Development*. London: Zed, 1989.

Sibley, Mulford Q. "The Relevance of Classical Political Theory for Economy, Technology & Ecology," *Alternatives* (Winter 1973): 14–35.

————. *Nature and Civilization: Some Implications For Politics*. Itasca, Ill.: F.E. Peacock Publishers, 1977.

Sigmund, Paul E. *Natural Law in Political Thought*. Cambridge, Mass.: Winthrop, 1971.

Simon, Julian. *The Ultimate Resource*. Princeton, N.J.: Princeton University Press, 1982.

Skinner, Quentin. "The Context of Hobbes' Theory of Political Obligation," in *Hobbes and Rousseau: A Collection of Critical Essays*, Maurice Cranston and Richard Peters, eds. Garden City, N.Y.: Anchor Books, 1972.

Smiley, Marion. *Moral Responsibility and the Boundaries of Community: Power and Accountability from a Pragmatic Point of View*. Chicago: University of Chicago Press, 1992.

Snow, Donald. "The Pristine Silence of Leaving It All Alone," in *A Wolf in the Garden: The Land Rights Movement and the New Environmental Debate*, Philip D. Brick and R. McGreggor Cawley, eds. Lanham Md.: Rowman and Littlefield, 1996.

Sommerville, Johann P. *Thomas Hobbes: Political Ideas in Historical Context*. New York: St. Martin's Press, 1992.

Soper, Kate. *What Is Nature?: Culture, Politics and the non-Human*. London: Blackwell, 1995.

Sorell, Tom. "The Science in Hobbes's Politics," in *Perspectives on Thomas Hobbes*, G.A.J. Rogers and Alan Ryan, eds. Oxford: Clarendon Press, 1988.

Soulé, Michael E., and Gary Lease, ed. *Reinventing Nature? Responses to Postmodern Deconstruction*. Washington, D.C.: Island Press, 1995.

Spragens, Thomas A., Jr. "The Politics of Inertia and Gravitation: The Functions of Exemplar Paradigms in Social Thought," *Polity* (1973): 288–310.

————. *The Politics of Motion: The World of Thomas Hobbes*. Lexington: University Press of Kentucky, 1973.

Spretnek, Charlene, and Fritjof Capra. *Green Politics*. Santa Fe, N.M.: Bear, 1986.

Spring, David, and Eileen Spring, eds. *Ecology and Religion in History*. New York: Harper and Row, 1974.

Stegner, Wallace. "The Legacy of Aldo Leopold," in *Companion to* A Sand County Almanac, J. Baird Callicott, ed. Madison: University of Wisconsin Press, 1987.

Stone, Christopher D. *Earth and Other Ethics: The Case for Moral Pluralism*. New York: Harper & Row, 1987.

Strauss, Leo. *Persecution and the Art of Writing*. Glencoe, Ill.: Free Press, 1952.

———. *Political Philosophy of Hobbes: Its Basis and Its Genesis.* Translated by Elsa Sinclair. Chicago: University of Chicago Press, 1952.

———. *Natural Right and History.* Chicago: University of Chicago Press, 1953.

———. "On the Basis of Hobbes's Political Philosophy," in *What Is Political Philosophy?* Chicago: University of Chicago Press, 1988.

Sylvan, Richard (see also Routley, Richard). "A Critique of Deep Ecology," *Radical Philosophy* 40 (1985): 2–12; 41 (1985): 10–22.

Szasz, Andrew. *Ecopopulism: Toxic Waste and the Movement for Environmental Justice.* Minneapolis: University of Minnesota Press, 1994.

Taylor, Bob Pepperman. "Environmental Ethics and Political Theory," *Polity* 23 (Summer 1991): 567–583.

———. *Our Limits Transgressed: Environmental Political Thought in America.* Lawrence: University Press of Kansas, 1992.

Taylor, Bron Raymond, ed. *Ecological Resistance Movements.* Albany: State University of New York Press, 1995.

Taylor, C.C.W. "Politics," in *Cambridge Companion to Aristotle,* Jonathan Barnes, ed. Cambridge: Cambridge University Press, 1995.

Taylor, Charles. "The Politics of the Steady State," in *Beyond Industrial Growth,* Abraham Rotstein, ed. Toronto: University of Toronto Press, 1976.

———. "The Nature and Scope of Distributive Justice," in *Justice and Equality, Here and Now,* Frank Lucash, ed. Ithaca, N.Y.: Cornell University Press, 1986.

Taylor, Paul. *Respect for Nature: A Theory of Environmental Ethics.* Princeton, N.J.: Princeton University Press, 1986.

Thomas, Keith. *Man and the Natural World.* Harmondsworth, U.K.: Penguin, 1983.

Torgerson, Douglas. *The Promise of Green Politics: Environmentalism and the Public Sphere.* Durham, N.C.: Duke University Press, 1999.

Tricaud, François. "Hobbes's Conception of the State of Nature from 1640 to 1651: Evolution and Ambiguities," in *Perspectives on Thomas Hobbes,* G.A.J. Rogers and Alan Ryan, eds. Oxford: Clarendon Press, 1988.

Tuck, Richard. "Hobbes and Descartes," in *Perspectives on Thomas Hobbes,* G.A.J. Rogers and Alan Ryan, eds. Oxford: Clarendon Press, 1988.

———. "Optics and Sceptics: The Philosophical Foundations of Hobbes's Political Thought," in *Conscience and Casuistry in Early Modern Europe,* Edmund Leites, ed. Cambridge: Cambridge University Press, 1988.

———. *Hobbes.* New York: Oxford University Press, 1989.

Vogel, Stephen, *Against Nature: The Concept of Nature in Critical Theory.* Albany: State University of New York Press, 1996.

von Fritz, K., and E. Kapp. "The Development of Aristotle's Political Philosophy and the Concept of Nature," in *Articles on Aristotle,* Vol. 2, Jonathan Barnes, Malcolm Schofield, and Richard Sorabji, eds. London: Duckworth, 1975.

von Leyden, W. *Aristotle on Equality and Justice.* New York: St. Martin's Press, 1985.

Walker, K.J. "The Environmental Crisis: A Critique of Neo-Hobbesian Responses," *Polity* (Fall 1988): 67–81.

Walzer, Michael. *Spheres of Justice: A Defense of Pluralism and Equality.* New York: Basic Books, 1983.

———. *The Company of Critics: Social Criticism and Political Commitment in the Twentieth Century.* New York: Basic Books, 1988.

Warren, Karen J. "The Power and the Promise of Ecological Feminism," *Environmental Ethics* 12 (Summer 1990): 125–146.

Warrender, Howard. *The Political Philosophy of Hobbes: His Theory of Obligation.* Oxford: Clarendon Press, 1957.

Waterlow, Sarah. *Nature, Change, and Agency in Aristotle's* Physics: *A Philosophical Study.* Oxford: Clarendon Press, 1982.

Watkins, John W.N. *Hobbes's System of Ideas.* 2nd ed. Aldershot, U.K.: Gower, 1989.

Westfall, Richard S. *The Construction of Modern Science: Mechanisms and Mechanics.* Cambridge: Cambridge University Press, 1977.

Weston, Anthony. "Beyond Intrinsic Value: Pragmatism in Environmental Ethics." *Environmental Ethics* 7 (Winter 1985): 321–339.

———. "Before Environmental Ethics," *Environmental Ethics* 14 (Winter 1992): 321–338.

———. *Back to Earth: Tomorrow's Environmentalism.* Philadelphia: Temple University Press, 1994.

Westra, Laura. *An Environmental Proposal for Ethics: The Principle of Integrity.* Lanham, Md.: Rowman and Littlefield, 1994.

Westra, Laura, and Thomas M. Robinson, eds. *The Greeks and the Environment.* Lanham, Md.: Rowman and Littlefield, 1997.

White, Lynn, Jr. "The Historical Roots of Our Ecologic Crisis," *Science* 155 (1967): 1203–1208.

———. "Continuing the Conversation," in *Western Man and Environmental Ethics: Attitudes toward Nature and Technology,* Ian Barbour, ed. Reading, Mass.: Addison-Wesley, 1973.

Wieland, W. "The Problem of Teleology," in *Articles on Aristotle,* Vol. 1, Jonathan Barnes, Malcolm Schofield, and Richard Sorabji, eds. London: Duckworth, 1975.

Williams, Raymond. *Keywords: A Vocabulary of Culture and Society.* London: Fontana, 1976.

———. *Problems in Materialism and Culture: Selected Essays.* London: Verso, 1980.

———. "Ideas of Nature," chap. in *Problems in Materialism and Culture: Selected Essays.* London: Verso, 1980.

———. "Socialism and Ecology," chap. in *Resources of Hope: Culture, Democracy, Socialism.* London: Verso, 1989.

Wilson, Harlan. "Postmodernism, Authority, and Green Political Theory." Paper presented at annual meeting of the American Political Science Association, San Francisco, August 29–September 1, 1996.

Winner, Langdon. *Autonomous Technology: Technics-out-of-Control as a Theme in Political Thought.* Cambridge, Mass.: MIT Press, 1977.

———. *The Whale and the Reactor: A Search for Limits in an Age of High Technology.* Chicago: University of Chicago Press, 1986.

———. "The State of Nature Revisited," in *The Whale and the Reactor.* Chicago: University of Chicago Press, 1986.

Wolin, Sheldon, S. *Politics and Vision: Continuity and Innovation in Western Political Thought.* Boston: Little, Brown, 1960.

———. *Hobbes and the Epic Tradition of Political Theory.* Los Angeles: William Andrews Clark Memorial Library, University of California, 1970.

Woodbridge, Frederick J.E. *Aristotle's Vision of Nature.* New York: Columbia University Press, 1965.

Worster, Donald. *Nature's Economy: A History of Ecological Ideas.* Cambridge: Cambridge University Press, 1985.

———, ed. *The Ends of the Earth: Perspectives on Modern Environmental History.* Cambridge: Cambridge University Press, 1988.

———. "The Ecology of Order and Chaos," in *The Wealth of Nature.* Oxford: Oxford University Press, 1993.

Yack, Bernard. "Natural Right and Aristotle's Understanding of Justice," *Political Theory* 18 (1990): 216–237.

———. *The Problems of A Political Animal: Community, Justice, and Conflict in Aristotelian Political Thought.* Berkeley: University of California Press, 1993.

———. *The Fetishism of Modernities: Epochal Self-Consciousness in Contemporary Social and Political Thought.* Notre Dame, Ind.: University of Notre Dame Press, 1997.

Zimmerman, Michael E. *Contesting Earth's Future: Radical Ecology and Postmodernity.* Berkeley: University of California Press, 1994.

———. "The Threat of EcoFascism," *Social Theory and Practice* 21 (Summer 1995): 207–238.

Index